Kluwer Academic Publishers

P.O. Box 17, 3300 AA Dordrecht, The Netherlands

W0227379

Dear Reader

We would very much appreciate receiving your suggestions and criticisms on the *Plant Tissue Culture Manual*. They will be most helpful during our preparations for future supplements.

Would you please answer the questions listed below, and send your comments with any further suggestions you may have, to *Ir. A. Plaizier* at the above-mentioned address.

Thank you for your assistance!

A. Plaizier
Publisher

————————————————————————————————

PLANT TISSUE CULTURE MANUAL

1. What errors have you found? (list page numbers and describe mistakes)
2. What protocols do you find to be confusing or lacking in detail? (list chapter numbers and page numbers and describe problems)
3. What protocols do you feel should be replaced in future supplements with newer (better) methods?
4. What new topics or other material would you like to see included in future supplements?

Please print or type your answers in the space below, and continue overleaf.

Name: Date:

Address:

Plant Tissue Culture Manual
Supplement 5, May 1995

INSTRUCTIONS FOR SUPPLEMENT 5

Preliminary pages
Replace: Table of Contents

Section D
Add after Chapter D8:

Section E
Add:

Section G
Add:

Section H
Add:

End matter
Replace: Index

PLANT TISSUE CULTURE MANUAL SUPPLEMENT 5

PLANT TISSUE CULTURE MANUAL

Supplement 5

Edited by:

K. LINDSEY

Department of Botany, University of Leicester, U.K.

Kluwer Academic Publishers

Dordrecht / Boston / London

Library of Congress Cataloging-in-Publication Data

Plant tissue culture manual: fundamentals and applications / edited by K. Lindsey
   p.   cm.
  Includes bibliographical references and index.
  ISBN 0–7923–1115–9 (acid-free paper).
  1. Plant tissue culture—Laboratory manuals.   I. Lindsey, K.
QK725.P587   1991   90–26765
581′.0724—dc20

Manual
ISBN 0–7923–1115–9

Supplement 1
ISBN 0–7923–1319–4

Supplement 2
ISBN 0–7923–1516–2

Supplement 3
ISBN 0–7923–2027–1

Supplement 4
ISBN 0–7923–2883–3

Supplement 5
ISBN 0–7923–3319–5

Published by Kluwer Academic Publishers,
P.O. Box 17, 300 AA Dordrecht, The Netherlands.

Kluwer Academic Publishers incorporates
the publishing programmes of
D. Reidel, Martinus Nijhoff, Dr W. Junk and MTP Press.

Sold and distributed in the U.S.A. and Canada
by Kluwer Academic Publishers,
101 Philip Drive, Norwell, MA 02061, U.S.A.

In all other countries, sold and distributed
by Kluwer Academic Publishers Group,
P.O. Box 322, 3300 AH Dordrecht, The Netherlands.

Printed on acid-free paper

Contents

SECTION C: PROPAGATION & CONSERVATION OF GERMPLASM

8. Clonal propagation of eucalypts
   J.A. McComb, Perth, Australia

SECTION D: DIRECT GENE TRANSFER & PROTOPLAST FUSION

1. Gene transfer by particle bombardment
   T.M. Klein, S. Knowlton and R. Arentzen, Wilmington, USA

2. Transformation of pollen by particle bombardment
   D. Twell, T.M. Klein and S. McCormick, Albany, USA

3. Electrical fusion of protoplasts
   M.G.K. Jones, Murdoch, Australia

4. Cybrid production and selection
   E. Galun and D. Aviv, Rehovot, Israel

5. Fluorescence-activated analysis and sorting of protoplasts and somatic hybrids
   D.W. Galbraith, Lincoln, USA

6. RFLP analysis of organellar genomes in somatic hybrids
   E. Pehu, Helsinki, Finland

7. Isolation and uptake of plant nuclei
   P.K. Saxena and J. King, Saskatoon, Canada

8. *In situ* hybridization to plant metaphase chromosomes: Radioactive and non-radioactive detection of repetitive and low copy number genes
   J. Veuskens, S. Hinnisdaels & A. Mouras, Talence, France

*9. Chemical fusion of protoplasts
   P. Anthony, R. Marchant, N.W. Blackhall, J.B. Power, M.R. Davey, Nottingham, UK

SECTION E: REPRODUCTIVE TISSUES

1. *In vitro* fertilisation of maize
   E. Kranz, Hamburg, Germany

2. Endosperm culture
   S. Stirn and H.-J. Jacobsen, Bonn, Germany

3. Endosperm culture
   B.M. Johri and P.S. Srivastava, Delhi, India

4. Hybrid embryo rescue
   A. Agnihotri, New Delhi, India

---

* Included in Supplement 5.

---

* Included in Supplement 5.

Plant Tissue Culture Manual **D9**, 1–15, 1995.

Chemical fusion of protoplasts

P. ANTHONY, R. MARCHANT, N.W. BLACKHALL, J.B. POWER & M.R. DAVEY
Plant Genetic Manipulation Group, Department of Life Science, University of Nottingham, University Park, Nottingham, NG7 2RD, U.K.

Introduction

Somatic hybridisation of plants offers a novel method of circumventing sexual barriers in plant breeding. It not only permits the production of new hybrids between sexually incompatible genotypes [14], but also serves as a method for the genetic modification of vegetatively propagated crops, sterile or subfertile species and plants with relatively long life cycles [5], such as tree species. Somatic hybridisation involves four distinct stages; protoplast isolation, protoplast fusion, selection and plant regeneration followed by cytological and molecular analysis of regenerated plants.

The successful isolation of protoplasts is dependent on several factors such as tissue source (e.g. leaves, cell suspensions, cotyledons, roots, pollen tetrads), the plant species and cultivar, together with the enzyme mixture and the physiological status of the source material, including the nature of the cell wall. The overriding consideration is protoplast viability, not only after isolation, but also after subsequent fusion.

Fusion can be mediated by either chemical or electrical techniques [21; see also chapter D3, this volume]. In the case of chemical fusion, relatively high concentrations of fusogens such as sodium nitrate [17], calcium nitrate [22], polyvinyl alcohol [12], dextran sulphate [8] or polyethylene glycol (PEG) [9] have been employed, sometimes in combination with high pH/Ca^{2+} [1]. When in close contact, the protoplast plasma membranes are temporarily destabilized by electrical stimulation or the action of the fusogen, which results in pore formation and permits cytoplasmic linkage between adjacent protoplasts. These linkages are thought to inhibit pore closure and permit randomly orientated lipid molecules on the periphery of the pores to align and form membrane bridges between adjacent protoplasts [2]. Subsequent mixing of cytoplasms occurs, followed by nuclear fusion, or, more frequently, by common spindle formation during the first mitosis. There is evidence to suggest that the pretreatment of protoplasts with non-ionic surfactants, or their inclusion in the fusogen solution, may promote protoplast fusion.

Protoplast fusion results in the production of a mixed population of heterokaryons, homokaryons and unfused parental protoplasts. Under optimum culture conditions, heterokaryons undergo cell wall formation and mitosis, often resulting in somatic hybrid cells by days 2–4 of culture. It is frequently necessary to utilise a selection strategy, involving either a means of isolating the heterokaryons, or a genetic complementation system designed to preferentially select for hybrid cells. Methods for the selection of heterokaryons have employed techniques such as manual micromanipulation [10] or automated flow cytometry [6]. Approaches for the selection of hybrid cells include those based on resistance to antibiotics, herbicides and amino acid analogues, and complementation involving the use of iodoacetamide treatment [13], x-ray or γ irradiation [20], or albino mutants [19]. Hybrid cells must be totipotent and capable of development, via embryogenesis or organogenesis, into somatic hybrid plants. In practical terms, shoot regeneration capacity need only be established for one of the parental protoplast systems.

Somatic hybrid plants can be characterised at the morphological, cytological, biochemical and molecular levels. Routine biochemical characterisation includes analyses of isozymes and fraction 1 protein, and the resistance of plants to viral infection and fungal toxins, together with their sensitivity to herbicides and antibiotics. Genetic analysis can be performed, provided hybrid plants are fertile [18]. More recently, the molecular techniques of RFLP and RAPD analyses have been adopted [4, 15, 16], while flow cytometry provides a rapid method for analysing the nuclear DNA content for ploidy determinations [7].

Techniques for large-scale chemically induced protoplast fusion

The following procedures are routine, well proven methods for the isolation and chemically-induced fusion of leaf mesophyll protoplasts of *Petunia parodii* with cell suspension-derived protoplasts of *P. hybrida* [18]. They can be adapted readily to any combination of protoplast systems. In this model system, heterokaryons are easily identified as they contain chloroplasts from the leaf mesophyll-derived protoplast partner, in the highly cytoplasmic background which originates from the cell suspension-derived protoplasts. Such a visual marker system permits optimisation of fusion conditions, including the ratio of the parental protoplasts in the fusion mixture and the temperature and duration of exposure to the fusogen. Despite the cytotoxic effects of PEG to protoplasts of some species, a method using purified PEG, with a reduced carbonyl content, ensures both higher frequencies of heterokaryon formation and viability than can be obtained using unpurified PEG preparations [3]. Autoclaving of PEG causes an increase in the carbonyl content. Consequently, it is advisable to filter sterilise PEG solutions. Storage should be in the dark at 4 °C for as short a time as possible, ideally less than a week. Some manufacturers now supply purified PEG as a sterile solution protected from the light in a suitable container and under nitrogen. This is a particularly convenient way of handling PEG.

Procedures

Isolation of leaf mesophyll protoplasts of Petunia parodii

Steps in the procedure
1. Detach several fully expanded basal leaves, including petioles, from plants (approx. 40 days old from seed germination) and place, together with two disposable medical gloves, in a sterile casserole dish. Surface sterilise leaves and the gloves by immersion in 7.5% (v/v) "Domestos" bleach solution (or the equivalent hypochlorite solution with wetting agent) for 30 min.
2. Using one sterilised glove, transfer the leaves to a second casserole dish and rinse both the leaves and the gloves three times in sterile tap water. Decant the water between each rinse.
3. Place the leaves on a rigid sterile surface (e.g. a sterile white tile). Remove the lower epidermis of the leaves by peeling with fine forceps, whilst holding each leaf flat with a gloved hand; discard the epidermis. Place the peeled leaves, adaxial side downwards, on the surface of approx. 30 ml of CPW13M solution (see Solutions) contained in a 14 cm Petri dish; completely cover the liquid surface with peeled leaf explants, but avoid overlapping the peeled explants. When yield per g. fresh weight is required, the leaf pieces can be weighed by placing them in a pre-weighed Petri dish containing CPW13M solution.
4. Leave the peeled explants on CPW13M solution for 30 min to 1 h; pipette off the CPW13M solution and replace immediately with 25 ml of the enzyme solution (see Solutions).
5. Incubate the leaves stationary overnight at 28 °C, in the dark.
6. Following incubation, remove the enzyme solution with a pipette, taking care not to disturb the digested leaf pieces. Replace the enzyme mixture with 20 ml of CPW21S solution (see Solutions). Release the protoplasts into the sucrose solution by gently squeezing the leaf material with a Pasteur pipette against the side of the Petri dish.
7. Transfer, by Pasteur pipette, the protoplast enzyme-mixture to 16 ml screw-cap centrifuge tubes. Adjust the volume of the liquid in the tubes, as necessary, with CPW21S solution, and centrifuge at $100 \times g$ for 10 min. Pipette the floating protoplasts (now free of debris) from the surface of the liquid and transfer to a clean centrifuge tube.
8. Wash the protoplast suspension once with CPW13M solution by resuspension and centrifugation ($80 \times g$ for 10 min). Discard the supernatant and resuspend the protoplast pellet in 10 ml (or a known volume) of CPW13M solution for the estimation of total pro-

toplast yield or yield per g. fresh weight, if the latter is required. Count the protoplasts using a haemocytometer.
9. Pellet the protoplasts by centrifugation (80 × g for 10 min), remove the supernatant and resuspend the protoplasts at 2.0×10^5 ml^{-1} in MSP19M liquid culture medium.

Isolation of cell suspension protoplasts of albino Petunia hybrida

Steps in the procedure
1. Use a cell suspension 3–5 d after subculture, at which time the cell walls are relatively thin.
2. Allow the cells to settle to the bottom of a 250 ml Erlenmeyer flask, decant the medium and remove any residual liquid with a Pasteur pipette. Alternatively, cells can be harvested on a 64 μm nylon sieve and transferred into a 250 ml flask. Add 30 ml of enzyme solution to 5–10 g fresh weight of cells (see Solutions) and incubate overnight (25 °C) at 40 rpm (horizontal shaker) in the dark.
3. Transfer, by pouring, the digested cells to 16 ml screw-capped tubes (top up with CPW13M solution as required) and centrifuge (80 × g for 10 min). Discard the supernatants and resuspend the pellets, containing protoplasts, in 5 ml per tube of CPW21S solution.
4. Pour the contents of 2–3 centrifuge tubes through a sterile 64 μm nylon sieve placed in the base of a 9 cm diameter Petri dish. Wash the sieve with 1–2 ml aliquots of CPW21S solution (several changes) until all the released protoplasts have passed through the sieve.
5. Transfer the filtrate (with protoplasts) to 16 ml screw-capped tubes (top up with CPW21S solution) and centrifuge (100 × g for 10 min).
6. Collect the floating protoplasts from the surface of the liquid in each tube and pool the samples in one clean centrifuge tube.
7. Wash the protoplasts once with CPW13M solution by resuspension and centrifugation (80 × g for 10 min). Discard the supernatant and resuspend the pellet in 10 ml of CPW13M solution; estimate the protoplast yield using a haemocytometer.
8. Pellet the protoplasts by centrifugation (80 × g for 10 min), remove the supernatant and resuspend the protoplasts at 2.0 × 10^5 ml^{-1} in MSP19M liquid culture medium.

General procedure and experimental design for the chemical fusion of protoplasts (volumes/replicates will vary depending on yield when applied to other species combinations)

Steps in the procedure
1. Suspend 7.2×10^6 protoplasts of *P. parodii* and *P. hybrida*, each at a density of 2.0×10^5 ml^{-1}, in 36 ml of MSP19M liquid culture medium (see Solutions and reference 18).
2. Dispense 4 ml of each protoplast suspension into two separate 16 ml centrifuge tubes to act as viability controls. A further 8 ml of each preparation should be dispensed into two separate centrifuge tubes to act as self-fusion controls to monitor viability and potential cross-feeding following culture medium-based selection for hybrids [18]. Add 4 ml of each of the two protoplast suspensions to three additional centrifuge tubes for the fusion treatments. More fusion tubes can be set up if protoplast yield permits.
3. Centrifuge all the tubes (except the viability controls) at $100 \times g$ for 5 min and discard the supernatants. Gently resuspend the pelleted protoplasts in 0.5 ml aliquots of MSP19M liquid medium.
4. Add the fusogen as detailed in the individual fusion protocols.
5. Following fusion (see separate protocols) centrifuge the tubes ($80 \times g$ for 10 min) and discard the supernatants. Resuspend the pelleted protoplasts in 16 ml per tube of MSP19M liquid culture medium (protoplasts are now at a density of 1.0×10^5 ml^{-1}).
6. Prepare several 9 cm Petri dishes (at least 12), each with 8 ml of agar-solidified MSP19M medium. From the fusion tubes (mixed populations of protoplasts) dispense 8 ml of suspended protoplasts per dish onto the surface of the medium. This gives a final, overall plating density of 5×10^{-4} protoplasts ml^{-1} of medium. For other species combinations, this may not be optimum plating density and the final protoplast plating density may have to be adjusted accordingly. Additionally, other culture approaches can be adopted, such as the use of liquid medium alone, and the embedding of protoplasts in agar or agarose droplets or layers.
7. Dilute each of the unfused viability controls with a further 4 ml of MSP19M medium (density now 1.0×10^5 ml^{-1}; 8 ml) and plate over 8 ml of agar-solidified medium as described earlier. Strictly, these controls should be subjected to identical washing and centrifugation procedures as for the fusion treated samples.
8. For the self-fusion controls, following fusion treatment and resuspension in 16 ml per tube of MSP19M medium, one 8 ml aliquot should be dispensed directly onto the agar-solidified medium counterpart (one dish per species-species fusion viability controls), whilst the remaining 8 ml aliquots of each self-fused species (at a density of 1×10^5 ml^{-1}) are mixed (1 : 1) to give two, 8 ml aliquots

for plating (8 ml per dish) over the agar-solidified medium. These crucial controls monitor not only post-fusion viability but, in the context of selection, possible crossfeeding and/or reversion in the albino *P. hybrida* parent. In other selection strategies, which rely on inhibition of growth of one or both parental protoplasts and their respective homokaryons, this type of control is equally essential.

9. Seal all Petri dishes with Nescofilm and culture at 25°C under 50 μmol m^{-2} s^{-1} of continuous "Daylight" fluorescent illumination.

Fusion protocols

I. Large-scale chemical fusion of protoplasts using PEG

Steps in the procedure
1. Prepare protoplasts for fusion according to steps 1–3 of the general procedure above.
2. Centrifuge all the tubes (80 × g for 10 min), except the viability controls, and resuspend the protoplasts in 0.5 ml of CPW11M solution.
3. Add 1.0 ml of 30% (w/v) PEG 1500 (see Solutions) and incubate the tubes for 20 min at 4 °C.
4. Add 1.5 ml of hypotonic solution dropwise with gentle agitation, to ensure rapid mixing. Repeat this addition twice at 2 min intervals.
5. Add 4 ml of MSP19M solution and incubate at 4 °C for 2 h in the fridge.
6. Follow steps 5–9 of the general procedure.

II. High pH/Ca²⁺ fusion

Steps in the procedure
1. Follow steps 1–3 of the general procedure.
2. Add 4 ml of high pH/Ca^{2+} fusion solution (see Solutions) to each of the tubes, except the viability controls.
3. Incubate the tubes at 30 °C for 10 min.
4. Centrifuge the tubes (50 × g for 4 min) and wash the protoplast pellets with CPW9M solution by resuspension and centrifugation (100 × g for 10 min).
5. Follow steps 5–9 of the general procedure.

III. PEG/high pH fusion

Steps in the procedure
1. Follow steps 1–3 of the general procedure.
2. Add 2 ml of PEG solution to each of the tubes, except the viability controls.
3. After 10 min incubation (23 °C), add 8 ml of high pH/Ca^{2+} solution to each tube and mix gently.
4. Following incubation (under identical conditions) for a further 10 min, centrifuge the tubes (50 × g for 4 min) and discard the supernatants. Wash the protoplast pellets once with CPW11M solution.
5. Follow steps 5–9 of the general procedure.

IV. Small-scale chemical fusion of protoplasts using PEG (for low yielding protoplast systems)

Steps in the procedure

1. Adjust the two parental protoplast suspensions to a density of 2.5×10^5 ml^{-1} in CPW13M solution. Mix equal volumes of the two protoplast suspensions in a centrifuge tube, using a minimum of 2 ml of each preparation.
2. Place a small drop (200 μl) of the protoplast mixture on a sterile coverslip contained in the bottom of a 5 cm Petri dish. Replace the lid and allow the protoplasts to settle for 5–10 min.
3. Dispense two 200 μl drops of 22.2% (w/v) PEG solution (see Solutions) on to the coverslip on each side of the protoplast mixture; allow the drops to coalesce with the CPW13M solution covering the settled protoplasts.
4. Leave the protoplasts undisturbed for 20–25 min, after which 200 μl of fusion solution is withdrawn and replaced with the same volume of Wash solution (see Solutions). Repeat this process at 5 min intervals over a 20 min period with a minimum of disturbance to the settled protoplasts.
5. Following elution of the fusion solution, replace the Wash solution with an equal volume of liquid MSP19M liquid medium (or alternative culture medium for other protoplast systems). Wash the protoplasts a minimum of three times, as in step (4).
6. Add a further 200 μl of liquid MSP19M liquid medium to give a final volume of 800 μl. Dispense the fusion-treated protoplasts in 100–200 μl drops in the base of the Petri dish. Seal the Petri dish with Nescofilm and culture at 25 °C under 50 μmol m^{-2} s^{-1} of continuous "Daylight" fluorescent illumination (or an appropriate light régime).

Notes

The selection scheme for this combination is based on differential parental protoplast growth responses to MSP19M medium. In this medium, *P. hybrida* protoplasts fail to divide, whereas *P. parodii* protoplasts form slow growing micro-calli. Hybrid cells form faster growing colonies due to heterosis; these colonies are easily identified.

Plant regeneration from somatic hybrid tissues, following chemical fusion and culture, can be achieved by the transfer (after 2–3 months) of protoplast-derived colonies onto agar-solidified MSZ medium (see Solutions). The resulting shoots are excised from the callus and rooted on agar-solidified MS-based medium lacking growth regulators. Rooted shoots are transferred to the glasshouse for *ex vitro* acclimation.

Solutions and culture media

– *CPW salts solution*
  1.48 g l^{-1} CaCl$_2$ · 2H$_2$O
  27.20 mg l^{-1} KH$_2$PO$_4$
  101.0 mg l^{-1} KNO$_3$
  246.0 mg l^{-1} MgSO$_4$ · 7H$_2$O
  0.16 mg l^{-1} KI
  0.025 mg l^{-1} CuSO$_4$ · 5H$_2$O
  CPW9M, CPW11M, CPW13M and CPW21S are CPW salt solutions, with 9%, 11%, 13% (w/v) mannitol or 21% (w/v) sucrose respectively. Adjust the pH to 5.8.

– *Enzyme solution for protoplast isolation from leaves of* P. parodii
  1.5% (w/v) Meicelase.
  0.05% (w/v) Macerozyme R10.
  400 mg l^{-1} Ampicillin
  10 mg l^{-1} Gentamycin
  10 mg l^{-1} Tetracycline
  CPW13M solution, pH 5.8

– *Enzyme solution for the isolation of protoplasts from albino cell suspensions of* P. hybrida
  2.0% (w/v) Rhozyme HP 150
  2.0% (w/v) Meicelase
  0.03% (w/v) Macerozyme R10
  CPW13M solution, pH 5.8

– *PEG solution for large-scale protoplast fusion*
  30% (w/v) low carbonyl content PEG (Mol. Wt. 1500) in pH 8.0 buffer (19.5 g l^{-1} HEPES sodium salt)

– *Hypotonic solution*
  90 g l^{-1} mannitol
  2.0 g l^{-1} bovine serum albumin
  Adjust to pH 5.8 and filter sterilise

– *High pH/Ca^{2+} solution*
  0.05 M glycine-NaOH buffer
  90 g l^{-1} mannitol
  11 g l^{-1} CaCl$_2$.2H$_2$O
  pH 10.4
  Sterilise by filtration. Use only freshly made preparations.

- *PEG solution for small-scale protoplast fusion*
  22.2% (w/v) PEG (Mol. Wt. 1500)
  18.0 g l^{-1} sucrose
  1.54 g l^{-1} $CaCl_2.2H_2O$
  95.2 mg l^{-1} KH_2PO_4
  pH 5.8

- *Wash solution*
  110 g l^{-1} sucrose
  7.4 g l^{-1} $CaCl_2.2H_2O$
  3.75 g l^{-1} glycine
  pH 5.8

- *UM liquid culture medium for cell suspension cultures of*
  P. hybrida
  MS salts [9]
  30 g l^{-1} sucrose
  2.0 mg l^{-1} 2,4-dichlorophenoxyacetic acid
  0.25 mg l^{-1} kinetin
  9.9 mg l^{-1} thiamine HCl
  9.5 mg l^{-1} pyridoxine HCl
  4.5 mg l^{-1} nicotinic acid
  2.0 g l^{-1} casein enzymatic hydrolysate
  pH 5.8

- *MSP19M protoplast culture medium*
  MS salts [9]
  30 g l^{-1} sucrose
  90 g l^{-1} mannitol
  2.0 mg l^{-1} α-naphthaleneacetic acid
  0.5 mg l^{-1} 6-benzylaminopurine
  pH 5.8
  MSPI9M agar-solidified medium is as above, but with 0.8% (w/v)
  agar.

- *MSZ plant regeneration medium*
  MS salts [9]
  30 g l^{-1} sucrose
  1.0 mg l^{-1} zeatin
  8 g l^{-1} agar
  pH 5.8

Materials and suppliers

1. 'Domestos' bleach: Lever Bros. Ltd., Kingston-upon-Thames, UK.
2. Disposable gloves: Becton Dickinson, Ltd., Wembley, Middlesex, UK.
3. Petri dishes: Sterilin Ltd., Teddington, Middlesex, UK.
4. Centrifuge tubes: Corning Glass Works, Corning, New York, USA.
5. Nylon sieves: Wilson Sieves, 2 Long Acre, Common Lane, Hucknall, Nottingham, UK.
6. Nescofilm: Bando Chemical Ind. Ltd., Kobe, Japan.
7. Rhozyme HP 150: Röhm and Haas, Philadelphia, USA.
8. Meicelase: Meiji Seika Kaisha, Tokyo, Japan.
9. Macerozyme R10: Kinki Yakult, Nishinomiya, Japan.
10. PEG Mol. Wt. 1500: Boehringer Mannheim UK Ltd., Lewes, East Sussex, UK.

References

1. Ahuja PS, Laiq-ur-Rahman, Bhargava SC. Banerjee S (1993) Regeneration of somatic hybrid plants between *Atropa belladonna* L. and *Hyoscyamus muticus* L. Plant Sci 92: 91–98.
2. Boni LT, Hui SW (1987) The mechanism of polyethylene glycol-induced fusion in model membranes. In: Sowers AE (Ed.) Cell Fusion, pp 119–124, Plenum Press, New York.
3. Chand PK, Davey MR, Power JB, Cocking EC (1988) An improved procedure for protoplast fusion using polyethylene glycol. J Plant Physiol 133: 480–485.
4. Craig AL, Morrison I, Baird E, Waugh R, Coleman M, Davie P, Powell W (1994) Expression of reducing sugar accumulation in interspecific somatic hybrids of potato. Plant Cell Rep 13: 401–405.
5. Gleba YY, Sytnik KM (1984) Protoplast fusion and parasexual hybridisation of higher plants. In: Shoeman R (Ed.) Protoplast Fusion, pp 36–62, Springer-Verlag, Berlin.
6. Hammatt N, Lister A, Blackhall NW, Gartland J, Ghose TK, Gilmour DM, Power JB, Davey MR, Cocking EC (1990) Selection of plant heterokaryons from diverse origins by flow cytometry. Protoplasma 154: 34–44.
7. Hammatt N, Blackhall NW, Davey MR (1991) Variation in the DNA content of *Glycine* species. J Exp Bot 42: 659–665.
8. Kishinami I, Widholm F (1987) Auxotrophic complementation in intergeneric hybrid cells obtained by electrical and dextran-induced protoplast fusion. Plant Cell Physiol 28: 211–218.
9. Louzada ES, Grosser JW, Gmitter FG (1993) Intergeneric somatic hybridisation of sexually incompatible parents: *Citrus sinensis* and *Atalantia ceylanica*. Plant Cell Rep 12: 687–690.
10. Mendis MH, Power JB, Davey MR (1991) Somatic hybrids of the forage legumes *Medicago sativa* L. and *M. falcata* L. J Exp Bot 42: 1565–1573.
11. Murashige T, Skoog F (1962) A revised medium for rapid growth and bioassays with tobacco tissue cultures. Physiol Plant 15: 473–497.
12. Nagata T (1978) A novel cell fusion method of protoplasts by polyvinyl alcohol. Naturwissenschaften 65: 263–264.
13. Nakano M, Mii M (1993) Interspecific somatic hybridisation in *Dianthus*: selection of hybrids by the use of iodoacetamide inactivation and regeneration ability. Plant Sci 88: 203–208.
14. Patil RS, Latif M, Vaz FBD, Davey MR, Power JB (1993) Hybridisation, through culture of embryos and immature seeds, of a range of tomato cultivars with a tomato somatic hybrid (*Lycopersicon esculentum* (+) *L. peruvianum*): emergence of a possible new marker gene for tomato breeding. Plant Breed 111: 273–282.
15. Pehu E, Thomas M, Poutala T, Karp A, Jones MGK (1990) Species-specific sequences in the genus *Solanum* – identification, characterisation and application to study somatic hybrids of *Solanum brevidens* and *Solanum tuberosum*. Theor App Genet 80: 693–698.
16. Polgar Z, Preiszner J, Dudits D, Feher A (1993) Vigorous growth of fusion products allows highly efficient selection of interspecific potato somatic hybrids – molecular proofs. Plant Cell Rep 12: 399–402.
17. Power JB, Cummins SE, Cocking EC (1970) Fusion of isolated plant protoplasts. Nature 225: 1016–1018.
18. Power JB, Frearson EM, Haywood C, George D, Evans PK, Berry SF, Cocking EC (1976) Somatic hybridisation of *Petunia hybrida* and *P. parodii*. Nature 263: 500–502.
19. Schoenmakers HCH, Wolters AMA, Nobel EM, de Klein CMJ, Koornneef M (1993) Allotriploid somatic hybrids of diploid tomato (*Lycoperiscon esculentum* Mill.) and monoploid potato (*Solanum tuberosum* L). Theor Appl Genet 87: 328–336.
20. Schoenmakers HCH, Wolters AMA, de Haan A, Saiedi AK, Koornneef M (1994) Asymmetric somatic hybridisation between tomato (*Lycopersicon esculentum* Mill) and gamma-irradiated potato (*Solanum tuberosum* L.): a quantitative analysis. Theor Appl Genet 87: 713–720.
21. Stattmann M, Gerick E, Wenzel G (1994) Interspecific somatic hybrids between *Solanum khasianum* and *S. aculeatissimum* produced by electrofusion. Plant Cell Rep 13: 193–196.
22. Wang GR, Binding H (1993) Somatic hybridisation between *Senecio fuchsii* and *S. jacobaea*. Acta Hort 36: 315–320.

Plant Tissue Culture Manual **E5**, 1–20, 1995.

In vitro culture of *Brassica juncea* zygotic proembryo

CHUN-MING LIU[a,1], ZHI-HONG XU[a,2] & NAM-HAI CHUA[b]
[a]Institute of Molecular & Cell Biology, National University of Singapore, Singapore 0511, Singapore; [b]Laboratory of Plant Molecular Biology, The Rockefeller University, 1230 York Ave., New York, NY10021–6399, U.S.A.; [1]To whom correspondence should be addressed. Current address: Department of Applied Genetics, John Innes Centre, Norwich NR4 7UH, U.K.; [2]Current address: Shanghai Institute of Plant Physiology, Chinese Academy of Sciences, Shanghai, 200032, China

Introduction

One of the most important approaches to studying embryo development is through embryo manipulation, which depends critically on the availability of an *in vitro* system. Three systems may be used for the manipulation – cultivation of zygotic embryos [10, 11], somatic embryos [3, 4, 19, 20, 25] and pollen embryos [8]. The advantages of using zygotic embryos rather than embryos induced from the somatic cells and pollens are their uniform pattern formation and the presence of a suspensor, which can be used as a visual indicator for the polarity of a globular proembryo. Somatic embryogenesis at the early stage is often characterised by cell proliferation that bears little resemblance to the polarized and highly regulated cell division in zygotic embryogenesis [2].

Proembryos are globular and heart-shape embryos preceding cotyledon initiation. Compared to the torpedo-shape and cotyledonary stage embryos, proembryos are heterotrophic in nature, dependent on the nutrients supplied by ovules and the surrounding endosperm for growth and development. By growing the embryos outside the environment of the ovules, it is possible to identify their nutritional requirements essential for the continued growth, cell differentiation and morphogenesis, which are difficult to determine while the embryos are enclosed in the ovules. Ideally, we would expect to follow the progressive embryogenesis *in vitro* starting from the fertilisation of an egg. Unfortunately, this goal is hard to achieve, especially in the stage from zygote to embryo when comprising only a few cells [5].

Cultivation of plant zygotic embryos was started by Hannig, who used a simple medium to culture 2 mm embryos at the turn of this century [7]. A practical usage of culturing embryos at this stage was to rescue embryos from interspecific hybridisation. A procedure for embryo rescue was contributed by Agnihotri [1]. This technique has been applied to a wide-range of species and many interspecific and intergeneric hybrids have been obtained through embryo culture. The first success in proembryo culture was not achieved until 1941 by Van Overbeek *et al.*, who observed that the growth of 150 μm-long *Datura* embryos was dramatically promoted by the addition of non-autoclaved coconut water to the culture medium [24]. Osmotic pressure of the medium is another critical consideration for culturing young embryos. Rietsema *et al.* (1952) found that the younger the embryo excised, the higher the medium osmolarity required for embryo culture [18]. Based on these observations, Ranghavan and Torrey (1963) and Nostog and Smith (1963) devised culture media that allowed the development of early heart-shape staged embryos into normal plants *in vitro* [15, 16, 17]. The importance of osmolarity in proembryo culture was confirmed by the measurement of the osmotic pressure of the milieu in the embryo sac of *Phaseolus vulgaris*. In this plant the osmolarity of the endosperm liquid was 0.7 mol/l when the embryos were at the heart-shaped stage, which decreased to 0.5 mol/l at the late cotyledonary stage [21]. Another major success in attempts to culture proembryos was achieved by Monnier [12, 13], who designed a system in which two media with different compositions were placed in juxtaposition in a Petri dish, ensuring a continual variation in the composition of the medium for the cultured embryos with time. Using this system, Monnier was able to culture *Capsella* embryos larger than 50 μm, but smaller embryos were still unable to develop under these conditions. Taking into considerations the requirements of both the osmolarity and nutrition, we successfully cultured early globular embryos (35 μm, containing 8–36 cells) of *Brassica juncea* (Indian Mustard) with high efficiency and the mature embryos germinated into fertile plants [10]. The culture system includes a double-layer culture system and a highly nutritious medium. This technique has also been used successfully in elucidating the role of auxin polar transport in embryonic pattern formation [11].

There are two major reasons for using *B. juncea* as a material for zygotic embryo culture. The first is the presence of a long suspensor connecting the embryo proper with the ovule tissue (Fig. 1), which makes dissection easier. In our previous work, we have observed that damage to the suspensor has no obvious effect on subsequent embryo development in culture [10]. Secondly, the development of *B. juncea* embryos follows that of a typical crucifer type in which the morphogenesis pattern has been well documented [22, 23].

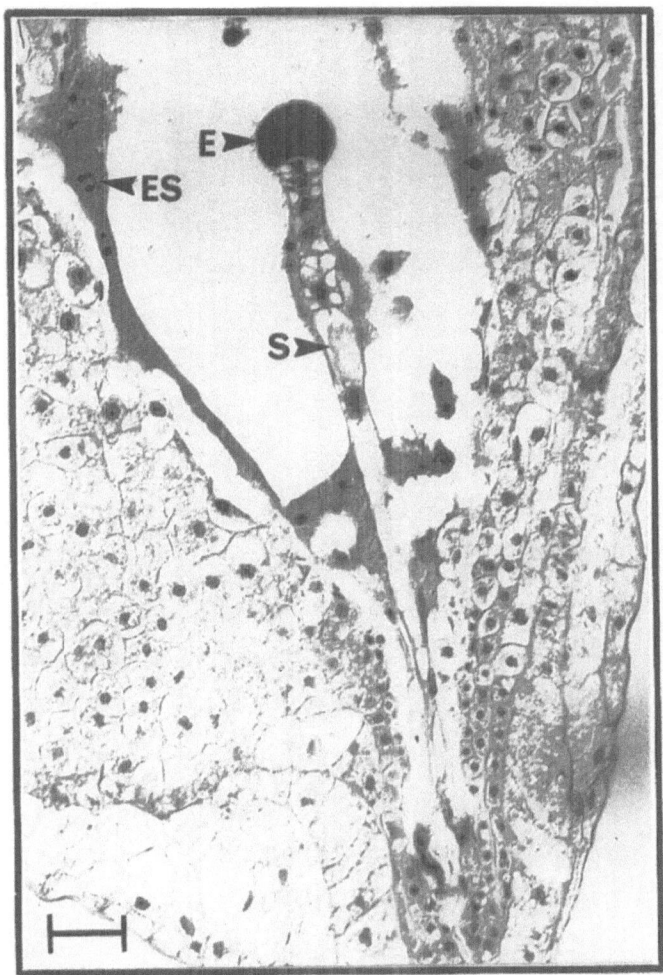

Fig. 1. Longitudinal section of a globular embryo of *B. juncea* in its ovule. The proembryo is connected to the maternal tissue through a long suspensor. Embryo dissection can be facilitated by holding the long suspensor, since damage to the suspensor has no evident effect on embryo development *in vitro*. E: embryo; S: suspensor; ES: endosperm. Bar = 50 μm.

Procedures

1. Major setups for embryo dissection

a) All dissection work is carried out in a sterile environment in an air-flow cabinet. A 30 W UV light mounted in the cabinet is used to sterilise the equipment such as the dissection microscope.
b) A dissection microscope. We used one from Nikon in which the light is supplied from the bottom of the dissection stage and which also has an adjustable reflection mirror. To observe the tiny embryo, we found it necessary to be able to adjust the angle of the mirror to achieve a dark-field effect.
c) A micropipette for embryo transfer, which is made by connecting a hand-drawn capillary pipette (ref. to this book, PTCM-A10/8) with a latex tube. The inner diameter of the open tip of the pipette should be 150–200 μm. The other end of the tube is blocked with a glass bead. The embryo can be sucked in and out by squeezing the tube (Fig. 2). The micropipette can be sterilised by immersing in

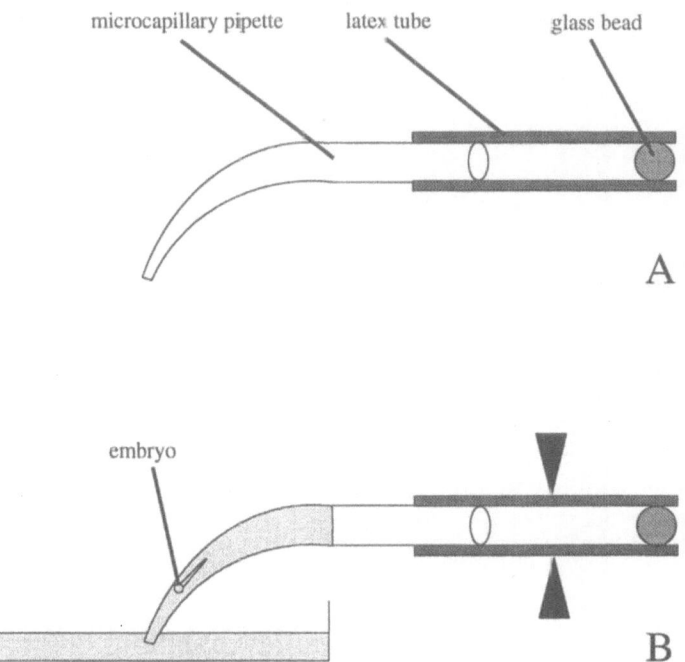

Fig. 2. Preparation of a micropipette for the embryo transfer. A. The micropipette is made by connecting a microcapillary pipette with a latex tube. The open end of the latex tube is blocked by a glass bead; B. Transfer the embryo by squeezing the tube. To prevent the embryo from sticking to the side wall of the micropipette, the micropipette should be filled up with the dissection solution (9% glucose) before sucking up the embryo.

70% ethanol for 10 min followed by air-drying in an air-flow cabinet.

2. Media preparations

The embryo culture medium (ECM, ref.10) is based on the KM8P's medium [9] with several modifications. Six stock solutions are made before preparing the medium.

Steps in the procedure
1. Prepare stock solutions

Stock 1 Macroelements (20X)

Components	Final concentrations	Weigh out for stock
NH_4NO_3	200 mg/l	2 g
KNO_3	1500 mg/l	15 g
$CaCl_2 \cdot 5H_2O$	850 mg/l	8.5 g
$MgSO_4 \cdot 7H_2O$	400 mg/l	4 g
$KH_2PO_4 \cdot 2H_2O$	100 mg/l	1 g
$Na_2EDTA \cdot 2H_2O$	37 mg/l	0.37 g
$FeSO_4 \cdot 7H_2O$	28 mg/l	0.28 g

Add distilled water to final volume 500 ml.

Stock 2 Microelements, as in B_5 medium [6], 1000X stock solution.

Stock 3 Sugar mixture (1000X)

Components	Final concentrations	Weigh out for stock
mannose	0.1 mg/l	10 mg
fructose	0.1 mg/l	10 mg
ribose	0.1 mg/l	10 mg
xylose	0.1 mg/l	10 mg
rhamnose	0.1 mg/l	10 mg
cellobiose	0.1 mg/l	10 mg
sorbitol	0.1 mg/l	10 mg

Add distilled water to final volume 100 ml.

Stock 4 Organic acids, as KM8P's [9], 100X stock solution (a powdered mixture is available from Sigma, K3004).

Stock 5 Vitamins and amino acids (200X)

Components	Final concentrations	Weigh out for stock
inositol	500 mg/l	25 g
glutamine	200 mg/l	10 g
thiamine · HCl	1 mg/l	50 mg
nicotinic acid	0.1 mg/l	5 mg
pyridoxine · HCl	0.1 mg/l	5 mg
d-biotin	0.01 mg/l	0.5 mg
casein hydrolysate	100 mg/l	5 g

Add distilled water to final volume 250 ml.

Stock 6   Coconut water
Final concentration is 300 ml/l.

2. Preparation of 2XECM

To prepare 200 ml medium, mix the appropriate amount of stock solutions as shown below, and adjust the volume first to 100 ml (2XECM), stir the mixture well.

stock 1	10 ml
stock 2	200 µl
stock 3	200 µl
stock 4	2 ml
stock 5	1 ml
stock 6	60 ml
sucrose	8 g
glucose	4 g

Add distilled water to final volume 100 ml, adjust the pH to 5.4.

3. Take 50 ml of the 2XECM solution, add 0.6 g agarose (low gelling temperature, SeaPlaque), dissolve by incubating in a water bath at 90 °C, and add distilled water to final volume 100 ml. This is the medium used for the bottom-layer in the two-layer culture system.
4. Sterilise the medium immediately by vacuum filtration using a 0.45 µm disposable filterware (Nalgene, U.S.A.).
5. Take the remaining 50 ml 2XECM solution, add 6 g sucrose, 0.6 g agarose (same as used in 2.3), dissolve by heating in a 90 °C water bath and then sterilise as in step 2.4. This medium will be used for the top-layer in the double-layer culture system.
6. Aliquot the media and store the aliquots at 4 °C. The media can be used for up to 2 months.

Notes
2.1. To make the stock solutions, all chemicals must be dissolved separately before mixing. In addition, $FeSO_4 \cdot 7H_2O$ and $Na_2EDTA \cdot 2H_2O$ have to be dissolved separately and mixed entirely before addition to the stock solution.
2.2. The coconut water is obtained from green coconuts bought from local markets in Singapore, which is filtered through a layer of Whatman No. 1 filter paper before sterilization by passing through a 0.45 µm filter (Nalgene, U.S.A.).
2.3. Stock solutions 3, 4, 5 and 6 should be stored in a −20 °C freezer to prevent contamination.
2.4. As measured by a vapour pressure osmometer, the osmotic pressure of the bottom-layer medium is 0.45 mol/l and the top-layer 0.63 mol/l.

3. Maintenance of plants

Plants of *B. juncea* are grown in a growth chamber under 14 hr light/10 hr dark at 24 °C day/20 °C night, respectively. The light intensity is about 4000 lux and relative humidity 70%. Under these conditions, the plants normally begin to flower in 2 months, and an additional 30 days are required to obtain mature seeds.

4. Selection of ovules and sterilisation

Steps in the procedure

1. Collect the siliques 6–7 days after anthesis. At this stage the length of the ovules varies from 1000 to 1200 μm.
2. Immerse the whole siliques in 70% (V/V) ethanol for 20 s and agitate by gentle swirling.
3. Wash twice with sterile water.
4. Surface-sterilise the siliques with 10% (V/V) commercial bleach (Clorox, sodium hypochlorite solution) for 10 min with shaking at 100 rpm.
5. Rinse four times with sterile water to remove all trace of bleach.

Notes

4.1. Based on our data comparing the size of the ovules with that of the embryos inside, we found that the size of the embryo at a particular developmental stage is quite constant and shows a good correlation to the ovule size. To obtain proembryos from 30 to 80 μm in length, ovules between 1000 and 1200 μm long are selected.

5. Embryo dissection and culture

To avoid osmotic shock to the proembryos, all the dissection work is carried out in a sterile 9% (W/V) glucose solution.

Steps in the procedure

1. Melt both the bottom- and top-layer ECM media gently in a microwave oven. Keep the melted top-layer medium in a 38 °C water bath for future use.
2. Transfer 300 µl bottom-layer of ECM medium to each well of a 24-well multiplate (Nuclon, Denmark) and allow agar to solidify before proceeding.
3. Split the silique into two with a pair of forceps and a dissection needle. Pick up ovules of the appropriate size and suspend in a 9% glucose solution in a 6 cm plastic Petri dish (Nuclon, Denmark).
4. Cut the ovule with a sharp needle following the steps shown in the diagram (Fig. 3A). Normally a zygotic embryo can be seen along the cutting edges at this stage.
5. Isolate the embryo from the mother tissue with dissection needles by holding the ovule with the forceps.

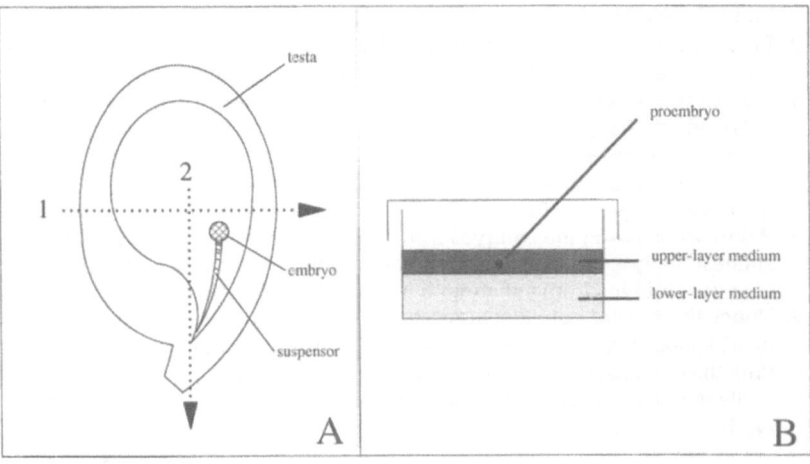

Fig. 3. A. A diagram shows the procedure for dissection of proembryo from *B. juncea*. Hold the ovule with a pair of sharp-point forceps, cut it with a dissection needle transversely, then cut the lower portion of the ovule longitudinally. Normally the embryo can be seen from the cut edge. Pick up the embryo by holding the suspensor with a pair of forceps; B. A diagram shows the double-layer culture system. 300 µl of ECM agarose medium is used for the lower-layer and a proembryo is embedded in 100 µl of upper-layer ECM medium which contains an additional 6% sucrose compared with the medium used in the lower layer.

6. Transfer one embryo onto the surface of the medium in each well of the multiplate using a micropipette (Fig. 2) and suck out the excess solution.
7. Continue dissection to obtain a suitable number of embryos for the experiments planned. In our work, 48 embryos were normally dissected for an experiment and 6 or 12 embryos were used for each treatment.
8. Transfer 100 μl of the top-layer medium (38 °C) to each well to embed the embryo (Fig. 3B).
9. Seal the plates with Nescofilm (Kobe, Japan).
10. Culture the plates in dark for the first 4 days in a 28 °C incubator, and then in a tissue culture room at the 1500 lux light for 18 hr each day, at the temperature of 28 °C/20 °C (day/night).
11. Observe the embryo development periodically with a Zeiss inverted phase-contrast microscope.

Notes

5.1. Tools for the dissection: a pair of sharp-point forceps (No. 4, Dumond, 72695-D, Switzerland); a home-made dissection needle assembled by connecting a sowing needle to a wooden handle.
5.2. As the proembryo is very fragile, it is important to avoid any direct mechanical damage to the embryo proper during the isolation. Great care should be taken not to directly touch the embryo proper with dissection needles or forceps. However, since damage of suspensor has no apparent effect on the subsequent embryo development in this species [10], the suspensor can be held with the forceps during embryo isolation.
5.3. Embryo culture plate: 24-well multi-plates (Nuclon) are used for the embryo culture. Plasticware from Nuclon was used because of its lower surface tension.
5.4. The culture system described above is suitable for culture proembryos larger than 35 μm in which 65% of 35–45 μm early globular- and 85% 45–60 μm globular stage embryos can develop into mature stages. Embryos larger than 80 μm at the early heart-shape stage grow very rapidly and develop into mature embryos at a frequency approaching 100%.
5.5. Before transferring the embryos from the dissection plate to the culture plate, we suggest firstly to fill up the micropipette with the dissection solution (Fig. 2) to prevent the embryos from adhering to the side wall of the micropipette.
5.6. Under the culture conditions described, a globular embryo will develop to the heart-shape stage after 2 days' culture, 6 days to the torpedo-shape stage (at this time the embryos can be seen by eye), and 8 days to the cotyledonary stage. Normally the embryos will reach the mature stage after 10 to 14 days in culture (Fig. 4A, B).

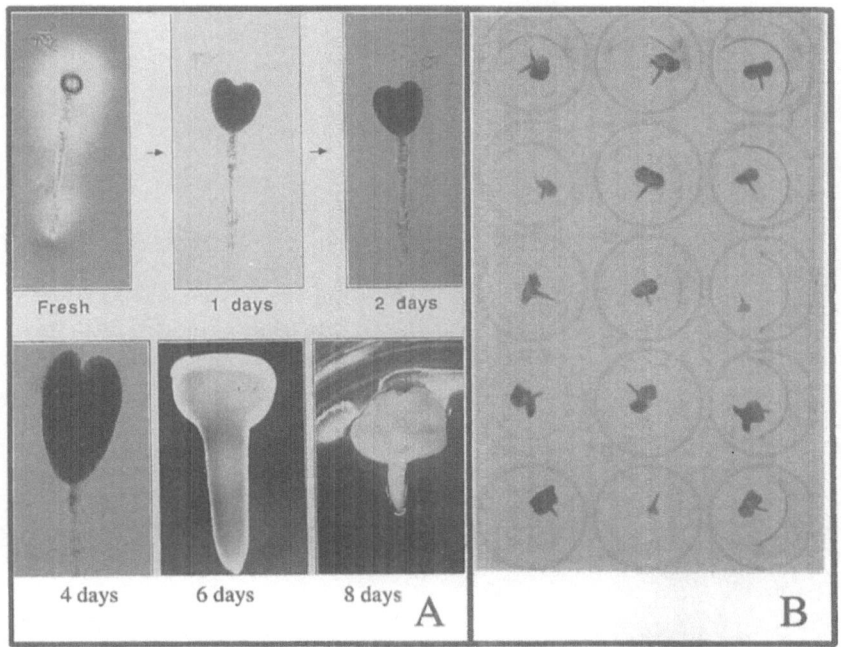

Fig. 4. Development of zygotic proembryos cf *B. juncea in vitro*. A. Sequential development of a globular embryo (45 μm in length) in culture. In most cases, a globular embryo develop to the cotyledonary stage after 8 days' culture; B. Part of a 24-well multiplate shows the mature embryos developed *in vitro* from the early globular stage after 10 days culture.

6. Germination of mature embryos

Germination of the mature embryo occurs only after transfer to fresh medium; failure to do so results in the mature embryos turning yellow and eventually dying. We found that the best medium for the embryo germination is the B_5 basal medium containing 1% sucrose. A higher sucrose content promotes precocious germination which results in abnormal plants.

Steps in the procedure
1. After 14–16 days culture in the initial medium (ECM), transfer the embryos onto solid B_5 basal medium containing 1% sucrose and 0.7% agar (extra pure bacteriological grade) by inserting the radical part of the embryos into the agar.
2. Seal the plate with Nescofilm and culture them in the tissue culture room for one month.
3. Transfer the seedlings into pots to obtain fertile plants.

Notes
6.1. It is important to insert the radicle of the mature embryo into the agar medium and avoid contact of the cotyledon with the culture medium.
6.2. Germination of the seedlings will be very slow in the first 10 to 15 days, which is necessary for obtaining normal plants. Precocious germination will cause the production of abnormal plants.

Acknowledgments

We would like to thank Drs. D.G. Neuhaus and B. Kost of Institut für Pflanzenwissenschaften, ETH Zurich, Switzerland, G.Y. Wang of Shanghai Institute of Plant Physiology, China for valuable discussion, and Dr. M. Leech for critical reading of the manuscript.

References

1. Agnihotri A (1993) Hybrid embryo rescue. In: Lindsey K (Ed.) Plant Tiss Cult Man E4, pp 1–8, Kluwer Academic, The Netherlands.
2. Carman JG (1990) Embryogenic cells in plant tissue cultures: occurrence and behaviour. In Vitro Cell Dev Biol 26: 746–753.
3. De Jong AJ, Cordewener J, Lo Schiavo F, Terzi M. Vanderckkove J, Van Kammen A, De Vries SC (1992) A carrot somatic embryo mutant is rescued by chitinase. Plant Cell 4: 425–433.
4. De Jong AJ, Heldstra R, Spaink HP, Hartog MV, Meijer EA, Hendriks T, Lo Schiavo F, Terzi M, Bisseling T, Van Kammen A, De Vries SC (1993) *Rhizobium* lipooligo-saccharides rescue a carrot somatic embryo mutant. Plant Cell 5: 615–620.
5. Dumas C, Mogensen HL (1993) Gametes and fertilisation: maize as a model system for experimental embryogenesis in flowering plants. Plant Cell 5: 1337–1348.
6. Gamborg OL, Miller RA, Ojima K (1968) Nutrient requirements of suspension cultures of soybean root cells. Exp Cell Res 50: 151–158.
7. Hannig E (1904) Physiology of plant embryos. I. The culture of cruciferous embryos outside the embryo sac. Bot. Ztg. 62: 46–81.
8. Huang B, Keller WA (1990) Microspore culture technology. J Tiss Cult Methods 12: 171–178.
9. Kao KN, Michayluk MR (1975) Nutrient requirements for growth of *Vicia hajastana* cells and protoplasts at very low population density in liquid medium. Planta 126: 105–110.
10. Liu CM, Xu ZH, Chua N-H (1993) Proembryo culture: *in vitro* development of early globular-stage zygotic embryos from *Brassica juncea*. Plant J 3: 291–300.
11. Liu CM, Xu ZH, Chua N-H (1993) Auxin polar transport is essential for the establishment of bilateral symmetry during early plant embryogenesis. Plant Cell 5: 621–630.
12. Monnier M (1976) Culture *in vitro* de l'embryon immature de *Capsella bursa-pastoris* Moench (l). Rev Cytol Biol Veg 39: 1–120.
13. Monnier M (1978) Culture of zygotic embryos. In: Thorpe TA (Ed.) Frontiers of Plant Tissue Culture, pp 277–286, University of Calgary Press, Calgary.
14. Murashige T, Skoog F (1962) A revised medium for rapid growth and bioassays with tobacco tissue cultures. Physiol Plant 15: 473–497.
15. Norstog K (1961) The growth and differentiation of cultured barley embryos. Am J Bot 48: 867–884.
16. Norstog K, Smith J (1963) Culture of small barley embryos on defined media. Science 142: 1655–1656.
17. Raghavan V, Torrey JG (1963) Growth and morphogenesis of globular and older embryos of *Capsella* in culture. Am J Bot 50: 540–551.
18. Rietsema J, Satina S, Blakeslee AF (1952) The effect of sucrose on the growth of *Datura stramonium* embryos *in vitro*. Am J Bot 40: 538–545.
19. Schiavone FM, Racusen RH (1990) Microsurgery reveals regional capabilities for pattern re-establishment in somatic carrot embryos. Dev Biol 141: 211–219.
20. Schiavone FM, Racusen RH (1991) Regeneration of the root pole in surgically transected carrot embryos occurs by position-dependent, proximodistal replacement of missing tissues. Development 113: 1305–1313.
21. Smith JG (1973) Embryo development in *Phaseolus vulgaris* II. Analysis of selected inorganic ions, ammonia, organic acids, amino acids and sugars in the endosperm liquid. Plant Physiol 51: 454–458.
22. Tykarska T (1976) Rape embryogenesis I. The proembryo development. Acta Soc Bot Pol 45: 3–16.
23. Tykarska T (1979) Rape embryogenesis II. Development of embryo proper. Acta Soc Bot Pol 48: 391–422.

24. Van Overbeek J, Conklin ME, Blakeslee AF (1941) Factors in coconut water milk essential for growth and development of very young *Datura* embryos. Science USA 94: 350–351.

25. Zimmerman JL (1993) Somatic embryogenesis: A model for early development in higher plants. Plant Cell 5: 1411–1423.

Plant Tissue Culture Manual **G3**, 1–23, 1995.
© 1995 *Kluwer Academic Publishers. Printed in the Netherlands.*

Biosynthesis of Monoterpene Indole Alkaloids *in vitro* [†]

WOLFGANG G.W. KURZ, FRIEDRICH CONSTABEL & ROBERT
T. TYLER*
*Plant Biotechnology Institute, National Research Council of Canada, Saskatoon, Saskatchewan, S7N
0W9, Canada; *Department of Applied Microbiology and Food Science, University of Saskatchewan,
Saskatoon, Saskatchewan, S7N 0W0, Canada*

Introduction

The medicinal value of monoterpene indole alkaloids continues to attract interest in their phytochemistry, chemical structure, biosynthesis, and chemothera-
peutic activity. Structural complexity and non-profitability of total synthesis,
perceived or real shortages in the supply of botanicals, and recent concerns for
the environmental impact of wild cropping, made production of these alkaloids
in vitro, i.e. by plant cells cultured in bioreactors, a viable prospect. The synthesis and accumulation of monoterpene indole alkaloids in cell cultures received
great attention when commercial production of compounds like quinine (1), ajmalicine (2) and catharanthine (6), vinblastine (8) and vincristine (9) appeared
attainable. Technologies aimed at increasing alkaloid accumulation by employment of production media, by precursor feeding, elicitation, semi-continuous
culture, and by application of enhanced bioreactors were designed to achieve
this goal. At present, however, efforts to further promote cell cultures as medici-
nals have softened somewhat due to regulatory barriers, but also due to biological barriers. Vinblastine and vincristine, the most desirable *Catharanthus* alkaloids, did not occur in concentrations which would warrant commercial exploi-
tation of an *in vitro* process. Research will be required to overcome these barriers. Research into the biosynthesis of monoterpene indole alkaloids, when
supplemented with molecular biological approaches, appears to be the most sen-
sitive approach.

Plant cell culture

The response of tissues excised from stems and leaves of *Catharanthus roseus*
(L.) G. Don to *in vitro* culture has been most favorable: crown gall tissue, ha-
bituated, and regular callus tissue have been grown *in vitro* since 1945. Culture
in bioreactors and occurrence of various alkaloids have been demonstrated as
early as 1969 [5]. DeLuca and Kurz [11] reviewed efforts to produce alkaloids
in *Catharanthus* cell cultures by one- and two-phase culture systems and produced a list of 30 plus alkaloids found. Single cell clones obtained with protoplasts derived from leaves showed extreme variation in alkaloid spectrum [7].
Cryopreservation allowed stability of such clones, potentially over several years

[†] NRC No. 38006.

[6]. Plant regeneration from cells [8], of importance for genetic engineering with this plant, may have gained in efficiency recently through embryogenesis in callus derived from anthers of specific germplasm, i.e. cv Little Delicata, using seeds from Takii & Comp., Tokyo [21].

Biosynthesis and production

Since the discovery of the hypoglycemic effects of *Catharanthus* alkaloids, but in particular since the demonstration of the antileukemic effect of vincaleucoblastine (VBL) in 1958, the structure of *Catharanthus* alkaloids has been under

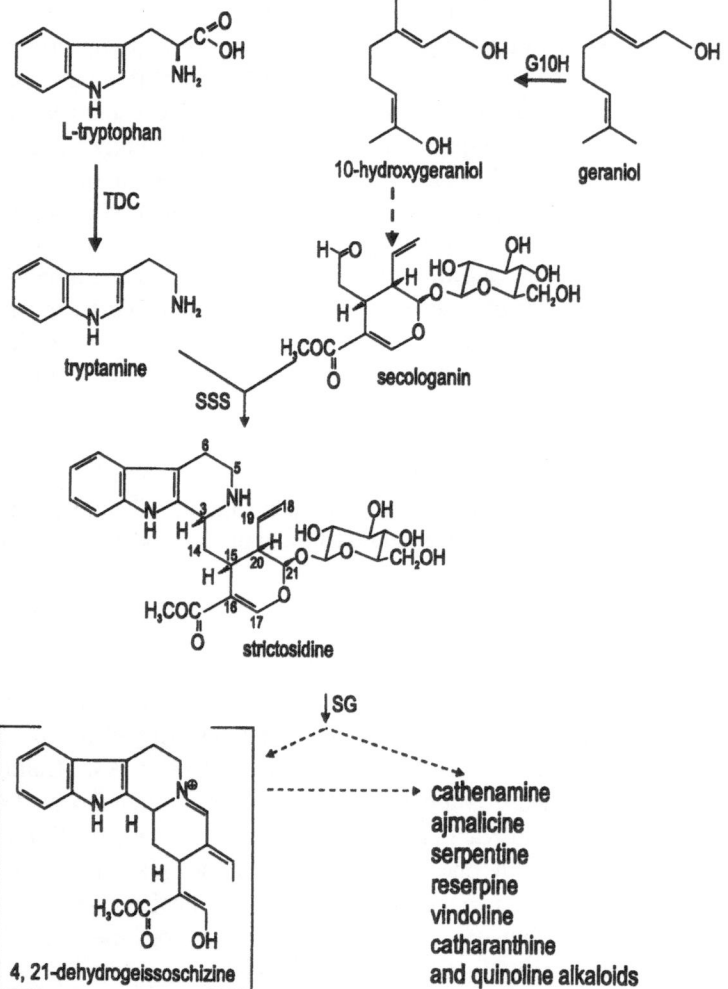

Fig; 1. Schematic representation of the biosynthesis of monoterpenoid indole and quinoline alkaloids (L.H. Stevens, 1994).

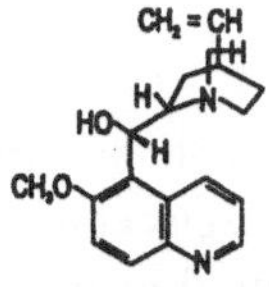

quine (1) 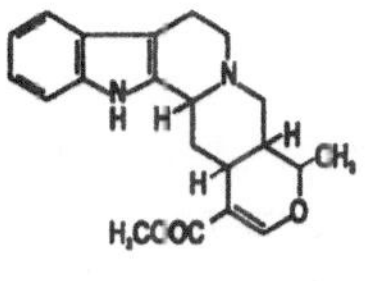 Ajmalicine (2) 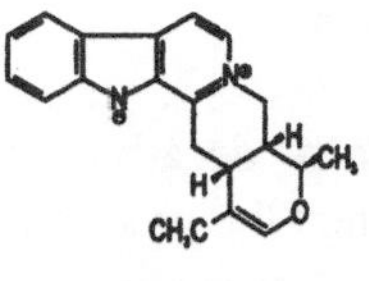 Serpentine (3)

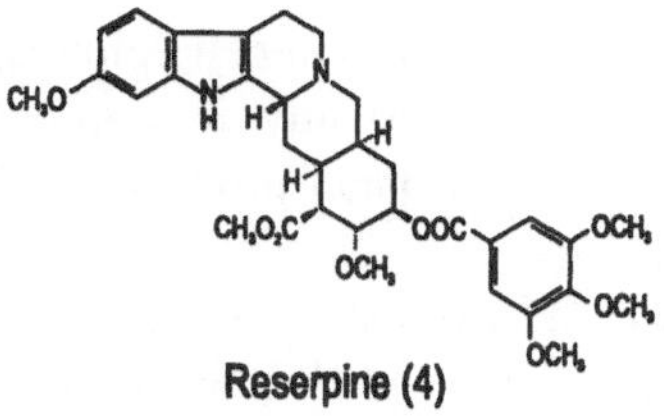

Reserpine (4) 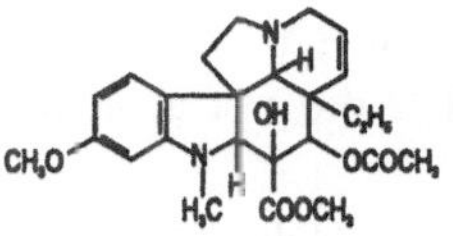 vindoline (5) 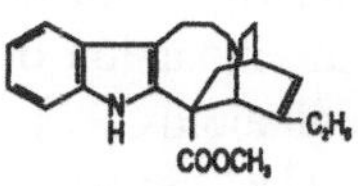 catharanthine (6)

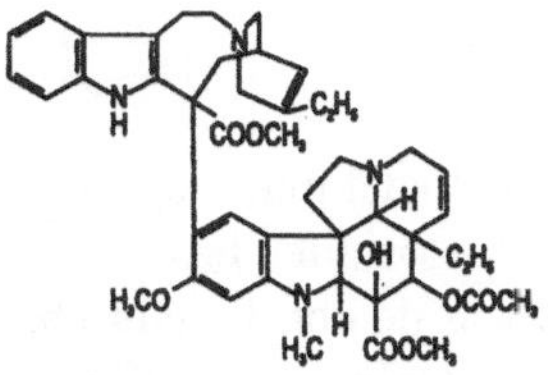

3', 4'-anhydrovinblastine (7) 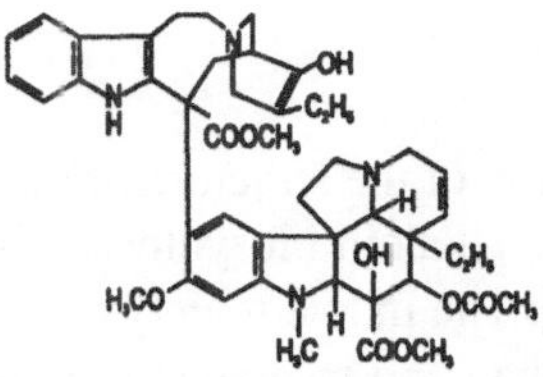 Vinblastine (8)

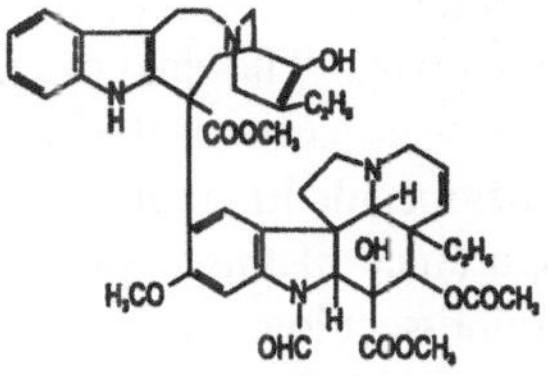

Vincristine (9)

Fig; 1. Continued

investigation [44]. Today, the number of compounds isolated and elucidated has exceeded 100 by far.

The biosynthesis, as presented in Fig. 1, has largely been investigated by using radio-labeled precursors. The focus has been on the first steps leading from tryptamine and secologanine to cathenamine and ajmalicine, the transformation of tryptophan/tryptamine to ajmalicine (2), serpentine (3), reserpine (4), vindoline (5), and catharanthine (6), and the condensation of catharanthine and vindoline to 3'4'-anhydrovinblastine (7) and vinblastine (8).

Tryptamine

The rate of decarboxylation of tryptophan to tryptamine was thought to critically affect the production of monoterpene indole alkaloids in *C. roseus* cell cultures. Determination of the tryptamine pool size, but more so the kinetics of tryptophan decarboxylase (TDC), were early targets. Interestingly, in *C. roseus* seedlings, the strictosidine synthase (SSS) enzyme activity appeared several days earlier in seedling development than TDC, and no traces of tryptamine could be found in young seedlings until the appearance of TDC activity. In cell cultures, TDC activity coincided with that of the SSS enzyme. Importantly, in cultured cells, the induction of TDC did not always result in subsequent production of indole alkaloids [12].

The TDC gene has been isolated, characterized, and expressed in transgenic tobacco plants, now showing varying levels of tryptamine [40]. The question remains whether similar transgenic *C. roseus* plants or cells might respond with varying levels of alkaloid production due to TDC gene over-expression.

Strictosidine

The elucidation of the structure of the first key alkaloidal intermediate in monoterpene indole alkaloid biosynthesis, 3-alpha(S)-strictosidine, has been reviewed recently [24]. The indole moiety of this alkaloid was demonstrated to be derived from tryptophan and tryptamine. Its monoterpene origin was confirmed by specific incorporation of geraniol and, subsequently, of loganine and secologanine in strictosidine and cathenamine/vindoline of *C. roseus* plants. Elucidation of strictosidine synthesis was corroborated by employing cell-free extracts obtained with *C. roseus* plants. The enzyme S-adenosyl-L-methionine:loganic acid methyl transferase was partially purified and characterized and, thus, marked the opening of enzyme identification along the indole alkaloid biosynthestic pathway. The importance of this approach was recognized through characterization of the tryptamine-secologanine coupling enzyme, dubbed strictosidine synthase (SSS). The gene for this enzyme has been cloned and expressed heterologously. Ajmalicine was the first alkaloid for which the biosynthesis was completely clarified at the enzyme level. Strictosidine synthase activity has been demonstrated for *Catharanthus roseus* and for *Cinchona* spec. cell cultures as well [47].

Ajmalicine (2)

Schübel *et al.* [37] demonstrated a 10-step pathway from strictosidine to ajmalicine. They found raucaffricine, a glycoalkaloid, as intermediate in ajmalicine biosynthesis. In *R. serpentina* cell cultures, this alkaloid occurred at concentrations of up to 1.6 g/l. *Rauvolfia* cell cultures have shown to be the best producers of ajmalicine, at levels up to 0.5% of dry weight.

Catharanthine (6)

Feeding radioactively labeled tryptophan to young plants of *Rauvolfia serpentina* Benth. and of *C. roseus* led to incorporation of label into ajmalicine (2), serpentine (3), reserpine (4), vindoline (5), and catharanthine (6). Increased accumulation of catharanthine was reported for cell cultures of *C. roseus* exposed to elicitors. Depending on the cell line, up to 5 and 30 mg/l of catharanthine was recorded with cells exposed to fungal elicitors and to vanadyl sulphate, respectively [14, 38]. Employment of cell lines selected for resistance to 5-methyltryptophan raised the catharanthine level in cells to over 200 mg/l [18].

Vindoline (5)

Vindoline has been shown to occur in *C. roseus* leaves, and appears not to accumulate in cell cultures. As long as it cannot be demonstrated for cells cultured *in vitro*, chances for synthesis of vinblastine through its coupling with catharanthine are dim. The elucidation of the pathway leading to vindoline was in part motivated by hopes to identify and rectify the enzymatic dysfunction in cultured cells. As a result, the synthesis of vindoline has become known to originate from tabersonine by a sequence: aromatic hydroxylation, O-methylation, hydration of 2,3-double bond, N(1)-methylation, hydroxylation at C-4, and 4-O-acetylation. All of these steps have been confirmed by isolation of the respective enzymes and determination of their activity [9, 10]. Studies with germinating seedlings have suggested that the three last steps are expressed in later developmental stages and the last two steps in light-grown seedlings only [1]. Strict developmental control may hinder vindoline synthesis under *in vitro* conditions.

Vinblastine (8)

Chemical coupling of catharanthine and vindoline has been accomplished by a modified Polonovski reaction in which 3'4'-anhydrovinblastine is formed. Biological coupling of the two monomers to vinblastine in *C. roseus* has been demonstrated for seedlings and with cell-free extracts of leaves. Finally, enzymatic conversion of anhydrovinblastine to vinblastine has been reported for cell-free extracts from cell suspension cultures. An economic process for dimer synthesis has yet to be developed. For example, enzymatic coupling of catharanthine and vindoline has been accomplished by catalysis using horseradish peroxidase; anhydrovinblastine yields of between 40 and 50% were obtained under optimized conditions [19]. Also, studies by Kutney [25] have shown that vindoline and catharanthine are being coupled to dimeric indole alkaloids by an enzyme complex which can be isolated from plant cell cultures and which can be immobilized.

The formation of vinblastine in leaves of multiple shoots of *C. roseus* cultured in vitro (15 μg/g dry weight) would confirm tissue and organ development as a prerequisite for vinblastine formation [29], and appears not to be a viable technological prospect.

Monoterpene indole alkaloids – analytical methods

Monoterpene indole alkaloids have been quantitatively determined in plant tissues or tissue/cell cultures from several genera and numerous species. *In vitro* studies have emphasized *Catharanthus roseus, Rauvolfia* spp. and *Tabernaemontana* spp. The various methodologies which have been employed seem generally applicable to any or all species/culture systems, although the presence of particular alkaloids or chlorophyll, the occurrence of alkaloids of interest at exceedingly low concentrations, or a requirement to screen large numbers of samples would influence the choice of methodology. Most assays of monoterpene indole alkaloids are best envisioned as consisting of four steps, namely extraction, purification, separation, and detection/identification. With some methods/species/cultures, however, one or more steps might be substantially reduced in significance, eliminated, or combined with other steps. Examples would include radioimmunoassay methods, where little or no purification of components is required (given suitably low cross-reactivities) and species/cultures accumulating only a single alkaloid in significant concentration, which would obviate the need for separation from similar compounds.

Extraction

The solvent of choice for extraction of monoterpene indole alkaloids from fresh or dried material has usually been methanol, either hot or cold, and with or without prior or simultaneous maceration or sonication of tissue [4, 13, 17, 23, 26, 27, 28, 31, 38, 46]. Other extraction solvents employed include ethanol [32], 70%-aqueous-ethanol [35], 80%-aqueous-ethanol [2], ethyl acetate [39], and methylene chloride [41].

Purification

Crude extracts of monoterpene indole alkaloids from plant tissues or tissue/cell cultures generally require partial purification or "clean-up" prior to separation and detection/identification of individual alkaloids. Typically, purification has taken advantage of the marked effect of pH on the solubility of alkaloids in polar organic solvents. Kurz and Constabel evaporated methanol extracts from *C. roseus* to dryness under vacuum [23]. Acid (1N HCl) extracts of the residue were washed with ethyl acetate, following which the pH of the acid fraction was adjusted to 10 prior to extraction with ethyl acetate. The ethyl acetate fraction was evaporated to dryness under vacuum and the residue taken up in a small volume of ethyl acetate in readiness for separation and identification of the alkaloids present. Smith *et al.* [38] and Loyola-Vargas *et al.* [27] employed more or less similar purification procedures with extracts from *C. roseus*, as did Asada and Shuler [4], Payne *et al.* [32] and Sasse *et al.* [35], who substituted methylene chloride for ethyl acetate, and Morris *et al.* [31], who used chloroform.

Purification of crude extracts may be simplified and made more rapid through the use of Sep-Pak C18 cartridges (Millipore Corporation, Bedford, MA) or similar solid phase extraction technology, as has been described by Morris *et al.* [31], van der Heijden [46] and Lee and Shuler [26]. An extract containing the alkaloid(s) of interest is applied to a miniature liquid chromatography column of appropriate chemistry. The alkaloids adsorb relatively strongly to the column, which allows impurities to be washed from the column prior to elution of the alkaloids with an appropriate solvent and subsequent column regeneration and reuse.

Separation and detection/identification

Although assays of total indole alkaloid content have been performed on occasion [32], plant tissues and tissue/cell cultures typically yield mixtures of monoterpene indole alkaloids which require chromatographic separation prior to detection/identification of individual components.

Thin-layer chromatography (TLC) has been widely employed as a separation technique, both as a preliminary screen prior to some other analytical method such as high pressure liquid chromatography (HPLC) and in TLC-based separation-quantitation methods in their own right. Typically, alkaloid mixtures in methanol, ethyl acetate or some other solvent are spotted on silica-gel-coated TLC plates containing a fluorescent indicator, which are then developed with an appropriate solvent mixture. The method of Kurz and Constabel [23] employs ethyl acetate:methanol (9 : 1, v/v), although many other combinations have been used successfully [4, 13, 26, 28, 30, 31, 36, 39]. Spots corresponding to individual alkaloids are then detected/identified on the basis of one or more of several characteristics, including their UV fluorescence, chromogenic reaction with ceric ammonium sulphate spray reagent [16] or Dragendorf's reagent and sodium nitrite [31], and R_f value. Rapid, direct quantitation on TLC is possible via fluorescence or absorbance densitometry scanning [13, 27, 36]. Alternatively, concentrations of individual alkaloids may be estimated visually by comparison to known concentrations of standards [26] or by spectrophotometric or other analysis of material scraped from TLC plates and eluted with an appropriate solvent.

HPLC appears to be the method most commonly employed for quantitation of monoterpene indole alkaloids. A number of systems have been described, most employing reverse-phase columns, a variety of mobile phases and gradients, and UV or fluorescence detection [4, 13, 22, 31, 38, 39, 46]. Gas-liquid chromatography (GLC) has not been widely applied to the analysis of monoterpene indole alkaloids, although its use for determination of ajmaline in *Rauvolfia vomitoria* has been reported [17].

Radioimmunoassay (RIA) methods have been applied to the quantitation of various constituents in plants and cell cultures, including determination of ajmalicine and serpentine in *C. roseus* [3, 48] and ajmaline in *Rauvolfia* spp. [2]. Stated advantages include precise measurements of very low concentrations of compounds, even in crude extracts, and an ability to be mechanized and applied to the analysis of hundreds of samples per day, making it suitable for large-scale screening of whole plants and cell cultures [3].

Enzymology of indole alkaloid synthesis *in vitro*

Enzymes involved in the biosynthetic pathways of indole alkaloids have been much studied in the past two decades. Due to the regulatory blocks associated with cell differentiation, however, only a limited number of enzymes involved in the synthesis of indole alkaloids could be studied in cell and tissue culture.

The synthesis of indole alkaloids involves the condensation of tryptamine and secologanine to strictosidine. Tryptamine, the indole part of these alkaloids, is formed by decarboxylation of L-tryptophan and catalyzed by tryptophan decarboxylase (TDC-E.C.4.1.1.28) while secologanine, the monoterpenoid part, is derived from geraniol. The subsequent condensation of tryptamine and secologanine is catalyzed by strictosidine synthase (SSS-E.C.4.3.3.2) [30, 45]. Roewer *et al.* [34] found that both TDC and SSS were regulated in a coordinated manner. The following step in the biosynthetic pathway of indole alkaloids involves the removal of the glucose moiety from strictosidine catalyzed by strictosidine-β-D-glucosidase (SG), which has been found to be specific to Apocynaceae producing indole alkaloids [20]. It is believed that the resulting dialdehyde undergoes several intramolecular rearrangements resulting in the formation of 4,21-dehydrogeissoschizine [43], although this compound has so far not been isolated from any *in vitro* bioreaction.

Procedures

Enzyme extraction for TDC and SSS assays [15]

Steps in the procedure
1. Thaw cells (0.3–1.2 g fresh weight) in 1.25 ml grinding buffer containing 0.1 M HEPES, 1 mM DTT and 5 mM EDTA and homogenize for 30–40 seconds with an Ultra-Turrax drive equipped with a 1–10 ml shaft.
2. Transfer the extract to 1.5 ml microfuge tubes and centrifuge for 3 min in a microcentrifuge.
3. Apply the clear supernatants (1 ml aliquots) to Pharmacia PD-10 columns (12.5 ml bed size) which have been pre-equilibrated with 20 mM HEPES, pH 7.6 and 1 mM DTT.
4. Collect protein samples for enzyme assays in 2 ml portions.

Assays for TDC and SSS [*15*]

The assays for both tryptophan decarboxylase (TDC) and strictosidine synthase (SSS) activities follow basically the ones described by Sasse *et al.* [36] and Mizukami *et al.* [30], respectively.

TDC

Steps in the procedure
1. The reaction mixture for the TDC assay includes 50 nM L-tryptophan containing 2.1×10^{-3} MBq L-[methylene -^{14}C] tryptophan, 4 nM pyridoxal phosphate, 5 μM HEPES, pH 7.5 and enzyme in a total volume of 100 μl.
2. Incubate the mixture for 30 min at 30 °C and stop the reaction by basifying to pH 10 with 10% K_2CO_3.
3. Extract the tryptamine with ethyl acetate and separate by TLC on silica gel using a mixture of $CHCl_3$: MeOH: 25% NH_3 (5 : 4:1) and measure its concentration either directly at 275 nm or, after spraying the plates with 0.1% ninhydrin solution, at 395 nm using a TLC-scanner.
4. Prepare a calibration curve for tryptamine.

SSS

Steps in the procedure
1. The reaction mixture for the SSS assay includes 85 nM tryptamine containing 4.6×10^{-3} MBq tryptamine [side chain-2-^{14}C], 2.8 μM D-gluconic acid lactone, 1.3 μM secologanine, 5 μM HEPES, pH 7.5 and enzyme in a total volume of 120 μl.
2. Incubate the mixture for 30 min at 30 °C and stop by basifying to pH 10 with 10% K_2CO_3.
3. Extract the strictosidine by using three 2 ml portions of ethylacetate.
4. Combine the portions and evaporate.
5. Analyze the extract by thin-layer chromatography on silica gel G using acetone: MeOH: diethylamine (7 : 2:1) for development.
6. Scrape off the band corresponding to strictosidine and count in a scintillation counter.

Enzyme extraction for SG [33]

1. Homogenize washed and frozen cells for 1 min in a Warring Blender at maximum rpm for 1 min.
2. Determine the weight of the frozen cell powder.
3. Add to the frozen cell powder, per gram cell fresh weights, 0.05 g polyvinylpyrrolidone and 1 ml of 0.1 M Tris buffer, pH 8.0 containing 3 mM DTT and 1 mM EDTA.
4. Thaw the material and remove cell debris by centrifugation for 30 min at $10,000 \times g$.
5. Dissolve the pellet in 0.1 M NaH_2PO_4 buffer, pH 6.8, containing 1 mM DTT and 0.02% sodium azide.
6. Desalt this preparation with Sephadex G-25 equilibrated with the same buffer.

Assay for SG [42]

Steps in the procedure
1. Make up the reaction mixture to a total volume of 100 μl, by combining 25 μl of enzyme preparation with 0.625 mM of strictosidine dissolved in 0.1 M NaH_2PO_4 buffer, pH 6.3.
2. Incubate for 60 min at 30 °C.
3. Stop the reaction by adding 100 μl of 5% trichloroacetic acid.
4. Add 20 μl of internal standard (8 mM Codeine HCl) and clarify and analyze by HPLC using an 8 μl flow cell and a wavelength of 280 nm.
5. For the analysis, use a LiChrosorb RP-8 select B column, 7-μm particle size with an inner dimension of 4 mm and a length of 250 mm.
6. Analyse at room temperature at a flow rate of 1 ml/min.
7. Elute with a filtered and degassed mixture of 7 mM sodium dodecyl sulfate and 25 mM NaH_2PO_4 in methanol: water (68 : 32 v/v) pH 6.2.

References

1. Aerts RJ, DeLuca V (1992) Phytochrome is involved in the light regulation of vindoline biosynthesis in *Catharanthus*. Plant Physiol 100: 1029–1032.
2. Arens H, Deus-Neumann B, Zenk MH (1987) Radioimmunoassay for the quantitative determination of ajmaline. Planta medica 53: 179–183.
3. Arens H, Stöckigt J, Weiler EW, Zenk MH (1978) Radioimmunoassay for the determination of the indole alkaloids ajmalicine and serpentine in planta. Planta medica 34: 37–46
4. Asada M, Shuler ML (1989) Stimulation of ajmalicine production and secretion from *Catharanthus roseus*: effects of absorption in situ, elicitors and alginate immobilization. Appl Microbiol Biotechnol 30: 475–481.
5. Carew DP (1975) Tissue culture studies of *Catharanthus roseus*. In: Taylor WI, Farnsworth NR (Eds.) The *Catharanthus* Alkaloids, pp 193–208, Marcel Dekker Inc, New York.
6. Chen THH, Kartha KK, Kurz WGW, Chatson KB, Constabel F (1984) Cryopreservation of alkaloid-producing cell cultures of periwinkle (*Catharanthus roseus*). Plant Physiol 75: 726–731.
7. Constabel F, Rambold S, Chatson KB, Kurz WGW, Kutney JP (1981) Alkaloid production in *Catharanthus roseus* (L.) G. Don. VI. Variation in alkaloid spectra of cell lines derived from one single leaf. Plant Cell Rep 1: 3–5.
8. Constabel F, Gaudet-LaPrairie P, Kurz WGW, Kutney JP (1982). Alkaloid production in *Catharanthus roseus* cell cultures. XII. Biosynthetic capacity of callus from original explants and regenerated shoots. Plant Cell Rep 1: 139–142.
9. De Carolis E., DeLuca V (1993). Purification to homogeneity and characterization of a 2-oxoglutarate dependent dioxygenase involved in vindoline biosynthesis in *Catharanthus roseus*. J Niol Chem 268: 5504–5511.
10. DeLuca V, Balsevich J, Tyler RT, Eilert U, Panchuk BD, Kurz WGW (1986) Biosynthesis of indole alkaloids: Developmental regulation of the biosynthetic pathway from tabersonine to vindoline in *Catharanthus roseus*. J Plant Physiol 125: 147–156.
11. DeLuca V, Kurz WGW (1988) Monoterpene indole alkaloids (*Catharanthus* Alkaloids) In: Constabel F, Vasil IK (Eds.) Cell Culture and Somatic Cell Genetics of Plants, Vol 5, Phytochemicals in Plant Cell Cultures, pp 385–402, Academic Press Inc, New York.
12. DeLuca V, Fernandez JA, Campbell D, Kurz WGW (1988) Developmental regulation of enzymes of indole alkaloid biosynthesis in *Catharanthus roseus*. Plant Physiol 86: 447–450.
13. Duez P, Chamart S, Vanhaelen M, Vanhaelen-Fastré R, Hanocq M, Molle L (1986) Comparison between high-performance thin-layer chromatography-densitometry and high-performance liquid chromatography for the determination of ajmaline, reserpine and rescinnamine in *Rauvolfia vomitoria* root bark. J Chrom 356: 334–340.
14. Eilert U, Constabel F, Kurz WGW (1986) Elicitor-stimulation of monoterpene indole alkaloid formation in suspension cultures of *Catharanthus roseus*. J Plant Physiol 126: 11–22.
15. Eilert U, DeLuca V, Constabel F, Kurz WGW (1987) Elicitor-mediated induction of tryptophan decarboxylase and strictosidine synthase activities in cell suspension cultures of *Catharanthus roseus*. Arch Biochem Biophys 254 491–497.
16. Farnsworth NR, Blomster RN, Damratoski D, Meer WA, Cammarato LV (1964) Studies on *Catharanthus* alkaloids. VI. Evaluation by means of thin-layer chromatography and ceric ammonium sulfate spray reagent. Lloydia 27: 302–314.
17. Forni GP (1979) Gas chromatographic determination of ajmaline in the bark of the root of *Rauvolfia vomitoria*. J Chrom 176: 129–133.
18. Fujita Y, Hara Y, Morimoto T, Misawa M (1990) Semisynthetic production of vinblastine involving cell cultures of *Catharanthus roseus* and chemical reaction. In: Nijkamp HJJ *et al*. (Eds.) Progress in Plant Cellular and Molecular Biology, pp 738–743, Kluwer Academic Publishers, Dordrecht.

19. Goodbody AE, Endo T, Vukovic J, Kutney JP, Choi LSL, Misawa M (1988) Enzymatic coupling of catharanthine and vindoline to form 3'4'-anhydrovinblastine by horseradish peroxidase. Planta Med 54: 136–140.

20. Hemscheidt T, Zenk MH (1980) Glucosidases involved in indole alkaloid biosynthesis of *Catharanthus roseus* cell cultures. FEBS Lett 110: 187–191.

21. Kim SW, Song NH, Jung KH, Kwak SS, Liu JR (1993) High frequency plant regeneration from anther-derived cell suspension cultures via somatic embryogenesis in *Catharanthus roseus*. Plant Cell Rep 13: 319–322

22. Kurz WGW (1984) Isolation and analysis of alkaloids. In: Vasil IK (Ed.) Cell Culture and Somatic Cell Genetics of Plants, Vol 1, pp 644–650, Academic Press, San Diego.

23. Kurz WGW, Constabel F (1982) Production and isolation of secondary metabolites. In: Wetter LR, Constabel F (Eds.) Plant Tissue Culture Methods, 2nd Revised Edition, pp 128–131, National Research Council of Canada, Ottawa.

24. Kutchan TM (1993) Strictosidine:from alkaloid to enzyme to gene. Phytochem 32: 493–506.

25. Kutney JP (1991) Plant Cell Cultures and synthetic chemistry: a potentially powerful route to complex natural products. Synlett 1: 11–19.

26. Lee CWT, Shuler ML (1991) Different shake flask closures alter gas phase composition and ajmalicine production in *Catharanthus roseus* suspensions. Biotechnology Tech 5: 173–178.

27. Loyola-Vargas VM, Méndez-Zeel M, Monforte-González M, de Lourdes Miranda-Ham M (1992) Serpentine accumulation during greening in normal and tumor tissues of *Catharanthus roseus*. J Plant Physiol 140: 213–217.

28. Merillon J-M, Chénieux JC, Rideau M (1983) Time course of growth, evolution of sugar-nitrogen metabolism and accumulation of alkaloids in a cell suspension of *Catharanthus roseus*. Planta medica 47: 169–176.

29. Miura Y, Hirata K, Kurano N, Miyamoto K, Uchida K (1988) Formation of vinblastine in multiple shoot culture of *Catharanthus roseus*. Planta Med: 54, 18–20.

30. Mizukami H, Nordlöv H, Lee SL, Scott AI (1979) Purification and properties of strictosidine synthetase (an enzyme condensing tryptamine and secologanin) from *Catharanthus roseus* culture cells. Biochem 18: 3760–3763.

31. Morris P, Scragg AH, Smart NJ, Stafford A (1985). Secondary product formation by cell suspension culture. In: Dixon RA (Ed.) Plant Cell Culture: A Practical Approach, pp 127–167, IRL Press, Oxford.

32. Payne GF, Payne NN, Shuler ML (1988) Bioreactor considerations for secondary metabolite production from plant cell tissue culture: indole alkaloids from *Catharanthus roseus*. Biotech Bioeng 31: 905–912.

33. Pennings EJM, van den Bosch RE, van der Heijden R, Stevens LH, Duine JA, Verpoorte R (1989). Assay of strictosidine synthase from plant cell cultures by high-performance liquid chromatography. Anal Biochem 176: 412–415.

34. Roewer IA, Cloutier N, Nessler CL, DeLuca V (1992) Transient induction of tryptophan decarboxylase (TDC) and strictosidine synthase (SS) genes in cell suspension cultures of *Catharanthus roseus*. Plant Cell Rep 11: 86–89.

35. Sasse F, Buchholz M, Berlin J (1983a) Site of action of growth inhibitory analogues in *Catharanthus roseus* cell suspension cultures. Z Naturforsch 36C: 910–915.

36. Sasse F, Buchholz M, Berlin J (1983b) Selection of cell lines of *Catharanthus roseus* with increased tryptophan decarboxylase activity. Z Naturforsch 38C: 916–922.

37. Schübel H, Ruyter CM, Stöckigt J (1989) Improved production of raucaffricine by cultivated *Rauvolfia* cells. Phytochem 28: 491–494.

38. Smith JI, Smart NJ, Misawa M, Kurz WGW, Tallevi SG, DiCosmo F (1987) Increased accumulation of indole alkaloids by some cell lines of *Catharanthus roseus* in response to addition of vanadyl sulphate. Plant Cell Rep 6: 142–145.

39. Smith JI, Amouzou E, Yamaguchi A, McLean S, DiCosmo F (1988) Peroxidase from bioreactor-cultivated *Catharanthus roseus* cell cultures mediates biosynthesis at α-3',4'-anhydrovinblastine. Biotech Appl Biochem 10 568–575.

40. Songstad DD, DeLuca V, Brisson N, Kurz WGW, Nessler CL (1990) High levels of tryptamine accumulation in transgenic tobacco expressing tryptophan decarboxylase. Plant Physiol 94: 1410–1413.

41. Stevens LH, Schripsema J, Pennings EJM, Verpoorte R (1992) Activities of enzymes involved in indole alkaloid biosynthesis in suspension cultures of *Catharanthus, Cinchona* and *Tabernaemontana* species. Plant Physiol Biochem 30: 675–681.

42. Stevens LH (1994) Formation and conversion of strictosidine in the biosynthesis of monoterpenoid indole and quinoline alkaloids. Thesis, University of Leiden, The Netherlands.

43. Stöckigt J (1980). The Biosynthesis of Heteroyohimbine-Type Alkaloids. In: Phillipson JD, Zenk MH (Eds.) Indole and biogenetically related alkaloids, pp 113–141, Academic Press, London.

44. Svoboda GH (1975) Introduction. In: Taylor WI, NR Farnsworth (Eds.) The *Catharanthus* Alkaloids, pp 1–7, Marcel Dekker Inc, New York.

45. Treimer JF, Zenk MH (1979) Purification and properties of strictosidine synthase, the key enzyme in indole alkaloid formation. Eur J Biochem 101: 225–233.

46. van der Heijden R, Lamping PJ, Out PP, Wijnsma R, Verpoorte R (1987) High performance liquid chromatographic determination of indole alkaloids in a suspension culture of *Tabernaemontana divaricata*. J Chrom 396: 287–295.

47. Verpoorte R, van der Heijden R, Schripsema J, Hoge JHC, ten Hoopen HJG (1993) Plant cell biotechnology for the production of alkaloids: present status and prospects. J Nat Prod 56: 186–207.

48. Zenk MH, El-Shagi H, Arens H, Stöckigt J, Weiler EW, Deus B (1977) Formation of the indole alkaloids serpentine and ajmalicine in cell suspension cultures of *Catharanthus roseus*. In: Barz W, Reinhard E, Zenk MH (Eds.) Plant Tissue Culture and its Biotechnological Applications, pp 27–43, Springer-Verlag, Berlin.

Abbreviations

TDC, tryptophan decarboxylase; SSS, strictosidine synthase; DTT, dithiothreitol; EDTA, ethylenediaminetetraacetic acid; HEPES, 4-(2-hydroxyethyl)-1-piperazineethanesulfonic acid; SG, strictosidine-β-D-glucosidase.

Plant Tissue Culture Manual **H1**, 1–30, 1995.
© 1995 *Kluwer Academic Publishers. Printed in the Netherlands.*

Establishment of Photoautrophic Cell Cultures

WOLFGANG HÜSEMANN

Institut für Biochemie und Biotechnologie der Pflanzen, Universität Münster, Hindenburgplatz 55, D-48143, Münster, Germany

Introduction

Under the appropriate light and nutritional conditions, *in vitro* cultured plant cells form chlorophyll, develop functional chloroplasts and thus gain photosynthetic competence.

The mode of nutrition is termed photoheterotrophic or photomixotrophic, if carbon and energy provision of light-grown chlorophyllous cells is mainly from an exogenous sugar in the culture medium, whereas during photoautotrophic growth, carbon and energy provision of the cells is exclusively by photosynthesis.

Sustained photoautotrophic growth of *in vitro* cultured plant cells, in the absence of an exogenous sugar but in the presence of CO_2-enriched air (1%–2% CO_2; v/v) has been achieved now for about 30 different plant species.

The culture technique has advanced to a stage that it is possible now to propagate plant cell cultures under photoautotrophic conditions as a callus on the surface of a nutrient agar or as cell suspensions in small glass flasks, in two-tier culture vessels, as well as in various bioreactor and continuous culture systems (For review see: 8, 20).

The ultimate goal in the work with photoautotrophic plant cell cultures is to eliminate all organic compounds from the medium and to reduce the level of CO_2 to atmospheric concentrations so that the cells are growing in a simple mineral salt medium with ambient CO_2 as the only source of carbon and true photoautotrophism is reached. This has been achieved now for cell suspension cultures of *Chenopodium rubrum* [11], *Dianthus caryophyllous* [15], *Arachis hypogaea* [6], though cell culture growth under atmospheric air is very low. For example, biomass increase in photoautotrophic cell suspensions from *Chenopodium rubrum* is 500–600% within 2 weeks under high CO_2 (2% (v/v) CO_2), but only about 30% increase in biomass under ambient CO_2 concentration.

So far, most photoautotrophic cell cultures have been established from C_3 plants. The only photoautotrophic cell cultures established from C_4-plants are the *Amaranthus* species [21], but without exhibiting a typical C_4 photosynthesis. On the other hand, photosynthetic differentiation leading to the expression of the crassulacean acid metabolism (CAM) as has been achieved for photoheterotrophically growing callus cultures of *Kalanchoe blossfeldiana* [1, 12].

Despite any progress in establishing plant cell cultures, some unresolved problems still exist. This is the complexity of interactions of nutritional, physiological, biochemical and molecular-biological factors including the responsible photoreceptor(s) in controlling chloroplast differentiation in cultured plant cells and the difficulty in inducing chlorophyll formation and chloroplast differentiation in morphologically unorganized cell cultures from the *Gramineae*. Finally, the question of why cultured plant cells in general need high CO_2 concentrations far above ambient air level (1%–2% CO_2; v/v) for sustained photoautotrophic growth has not been answered up to now.

Culture conditions favouring chloroplast development and photosynthetic competence of the cells

As a general rule, lowering the sugar content in the nutrient medium and simultaneously increasing the CO_2 partial pressure far above ambient air level will stimulate photosynthetic development. Under this selection pressure only cells with high photosynthetic capacities will survive. Careful selection of viable green cells during subculture will finally result in photoautotrophically growing cell cultures. However, the culture conditions required for the formation of rapidly growing, friable, nonorganized callus cultures may not always be favourable for the induction of greening. The factors that favour chloroplast development and photosynthetic competence of the cells will be explained briefly.

A. Culture medium

Most often the Murashige/Skoog medium [14], and to a lesser extend the B5-medium according to Gamborg [5] are used. Modifications in the composition of the nutrient media are mostly restricted to the use of growth regulators, sugar and other organic components such as vitamins and amino acids.

B. Phytohormones

In general, auxin concentrations in the culture medium should kept below 10^{-7} M or auxin activity should be reduced for example by the change from 2,4-dichlorophenoxyacetic acid (2,4-D) to naphthaleneacetic acid (NAA), because auxin concentration optimal for cell growth may suppress chlorophyll formation. The use of indoleacetic acid (IAA) should be avoided, because this phytohormone is readily destroyed in the light.

The demand of cytokinins for chlorophyll formation in cultured cells is species-specific. Kinetin can exert its beneficial effect on chlorophyll formation and photosynthetic capacity of the cells, because it can stimulate chloroplast replication and maturation without inducing cell division.

The inhibitory effect of ethylene on chlorophyll formation in cultured cells can partially be abolished by elevated CO_2-concentrations in the culture atmosphere.

C. Carbohydrates

Based on numerous experimental studies, we know about the inhibitory effect of exogenous sugars of chlorophyll formation and photosynthesis in cultured plant cells. Chlorophyll formation in cultured cells is permitted as sugar concentration in the batch-culture medium is depleted and the specific growth rate declined [2, 3]. The mechanism of this effect is poorly understood. The concentrations of readily utilized sugars such as glucose or sucrose should be reduced to about 1.0%–0.5%.

D. Light

White light from fluorescent tubes (for example: Osram L 36W/11, day light) or from mercury vapor lamps (Osram HQLR de Luxe, 85 W) is used at about 100–120 μmol m^{-2} s^{-1}.

Continuous illumination as well as light/dark cycles (for example: 16 h light/ 8 h darkness) can be used. As the process of greening and development of chloroplasts in cultured cells is specifically stimulated by the blue region of the visible spectrum [16, 17], in some cases it may be necessary to put the callus cultures exclusively under blue light (400–450 nm; Philips TLD 36W/18) to initiate chloroplast development.

Procedures

Selection and maintenance of photoheterotrophic cell cultures

Experimental trials have now to be undertaken to initiate or stimulate chlorophyll formation in cultured cells. In the author's experience, callus cells growing on the surface of a nutrient agar may be used in preference to suspension cultures. Plating of callus cells on agar medium in Petri dishes permits the separation of even small individual green parts of the callus during subculture. Once chlorophyllous callus cells have been sufficiently multiplied during several passages of subculture, a chlorophyllous suspension culture can be established.

Steps in the procedure
1. Place approximately 3–5 g cells into 50 ml liquid medium in a 200 ml Erlenmeyer flask, seal with aluminium foil and incubate on a gyrotory shaker (120 rpm; Certomat R, B. Braun Biotech International, Melsungen, Germany) in white light at 100–120 μmol $m^{-2} s^{-1}$ and 26 ± 1 °C.
2. Separate cells from early stationary growth phase from the culture medium by filtering on a glass filter (Schott D 2) or through a 0.5 mm nylon gauze.
3. Select by visual monitoring only the greenest cell groups or microcalluses for subculturing.

This process of serial selection has to be continued until a rapidly growing (300%–400% increase in biomass within 14 days) and finely dispersed chlorophyllous cell suspension (50–60 μg chlorophyll/g fresh weight; approximately 5–10 μg chlorophyll/10^6 cells) has been obtained, that can be used as the starting material for the selection of photoautotrophic cell lines.

Selection and growth of photoautotrophic cell cultures

Selection procedures

Major prerequisites for establishing photoautotrophic cell cultures are the presence and maintainance of a high chlorophyll content and photosynthetic competence of the cells even in the phase of active cell division. Usually, the change from the photoheterotrophic to the photoautotrophic mode of nutrition and cell growth is accompanied by a transient drastic reduction in the growth rate, in the chlorophyll content and in the viability of the cells. Therefore, rapidly growing (300–400% increase in fresh weight within 2 weeks) and highly chlorophyllous cell cultures (approximately 5–10 μg chlorophyll/10^6 cells) should be available for this selection procedure.

As a general rule, lowering the sugar content in the nutrient medium and simultaneously increasing the CO_2 partial pressure far above ambient air level (1%–2% CO_2; v/v) will stimulate photosynthetic development. Under this selection pressure only cells with high photosynthetic capacities will survive and take up photoautotrophic growth.

Establishing a photoautotrophic cell culture is a long-term process, involving several months of serial selection under the appropriate selection pressure. For example, it took more than 6 months to obtain photoautotrophically growing cell cultures from *Chenopodium rubrum* [10], *Lycopersicon esculentum* [18], *Glycine max* [7], *Mesembryanthemum crystallinum* (Hüsemann, unpublished).

Principally, 2 different methods are used for selecting cells capable of sustained photoautotrophic growth.

A. Callus induction under photoautotrophic conditions

Callus induction under photoautotrophic conditions was first reported by Yasuda *et al.* [23]. This technique is based on callus induction from explanted leaves in the light at elevated CO_2 concentrations on a sugar-free nutrient agar in the presence of phytohormones.

B. Sequential change of cultured cells from photoheterotrophic to photoautotrophic growth

Most often, photoautotrophic cell lines are selected from highly chlorophyllous, photoheterotrophically growing callus cells. During the transition of the cells from the photoheterotrophic to the photoautotrophic mode of nutrition and growth, sugar may be omitted at once or in several steps from the culture medium. The slow adjustment of the cells to reduced growth rates by gradually lowering the sugar content in the nutrient medium (2%–1%–0.5%–0.25%–0% sugar) in the presence of elevated CO_2-concentrations (1%–2% CO_2; v/v) is obviously more beneficial and even perhaps necessary for plant cell cultures to survive.

For some plant species it may be necessary to keep the oxygen level low and to prevent accumulation of volatile compounds in the culture atmosphere. In this case, the cell cultures must continuously be aerated with a gas mixture of known composition (Fig. 3). Otherwise, the cells can be grown in a closed culture system with CO_2 coming from a carbonate/bicarbonate-buffer (Fig. 1, 2).

The establishment of photoautotrophic cell suspensions from *Chenopodium rubrum* will be described as an example.

Steps in the procedure
1. Highly chlorophyllous and rapidly growing photoheterotrophic cell suspensions (mean chlorophyll content: 50–70 μg chlorophyll/g fresh weight; 5–10 μg chlorophyll/10^6 cells; about 300% increase in fresh weight/10 days) are the starting material for the selection of cells capable of photoautotrophic growth.
2. The two-tier culture vessel system is used with 2% (v/v) CO_2 coming from a 2 M $KHCO_3$/2 M K_2CO_3 buffer solution (Table 1; Fig. 4). Usually 50 ml of the carbonate buffer solution is used to establish the desired CO_2 concentration in the gaseous atmosphere of the culture vessel.
3. Transfer 2 to 4 grams of cells into 30 ml Murashige/Skoog medium supplemented with 10^{-7} to 5×10^{-8} M 2,4-D.
4. Lower the sucrose content of the culture medium stepwise from initially 2% sucrose via 1%, 0.5%, 0.25% to complete omission of sucrose.
5. Grow the cells under continuous light (100–120 μmol m^{-2} s^{-1}) on a gyrotory shaker (Certomat R, B. Braun Biotech International, Melsungen, Germany) at 120 rpm and 26 ± 1 °C.
6. Carefully monitor the cell culture visually during subculture to select highly chlorophyllous cells.

7. Mechanically isolate or separate by filtering through nylon gauze of different pore size (200–600 μm) small green cell groups and subsequently collect on a 100 μm close-meshed nylon gauze.
8. Only transfer the greenest cells into fresh culture medium.

Notes
3. The culture medium for *Chenopodium rubrum* cells contains the mineral salts and vitamins according to Murashige and Skoog [14] supplemented with 10^{-7} M 2,4-dichlorophenoxyacetic acid (2,4-D) and varying amounts of sucrose.
   The pH of the culture medium is adjusted to 5.8 prior to autoclaving.
   Sterilize separately by autoclaving the empty two-tier culture vessel (openings sealed with aluminium foil), the culture medium (30 ml, Erlenmeyer flask sealed with aluminium foil) and the carbonate buffer solution (50 ml–100 ml, screw-cap Duran glass flask).
   Nylon gauze is purchased from W. Babendererde, Hamburg, Germany.

CO_2-supply and composition of the culture atmosphere

As already pointed out, elevated CO_2 concentrations far above ambient air level (1%–2% CO_2, v/v) are a prerequisite for selecting and maintaining photoautotrophic plant cell cultures.

Two different methods can be used to increase CO_2 partial pressure in the culture atmosphere.

A. Supply of CO_2-enriched air in an open culture system

The cells are aerated with a gas mixture of known composition enriched in CO_2. By this procedure, the accumulation of volatile compounds like ethylene in the culture atmosphere can be prevented and the oxygen concentration can be kept at a desired low level to reduce photorespiration.

A gas mixture of known composition can be established by mixing together the appropriate volumes of CO_2, N_2 and O_2 using specifically designed gas flow-meters. Otherwise, the appropriate volumes of CO_2-free air (air can be kept CO_2-free by passing through a column of soda lime, carbosorb brand, 10–16 mesh) are mixed together with pure CO_2 to give the desired CO_2-concentration in the air.

For example, a gas flow rate of about 6 ml min^{-1} will sufficiently aerate 50 ml cell suspension growing in 200 ml Erlenmeyer flasks.

B. CO$_2$-supply from a bicarbonate/carbonate-buffer mixture

High CO$_2$-concentrations varying from 0.04% to 3% CO$_2$ (v/v) can be established by a 2 M KHCO$_3$/2M K$_2$CO$_3$ buffer solution [19] in the closed system of a two-tier culture vessel [9, 10].

Under these conditions the oxygen level in the closed culture unit will be adjusted by photosynthesis and respiration of the cells and volatile compounds may accumulate.

The CO$_2$ partial pressures above the buffer solutions can be calculated for 25 °C after the formula $(KHCO_3)_2/(K_2CO_3 \times CO_2) = K$. The value for $K_{25 °C}$ has been determined by interpolating between $K_{20 °C} = 3.35 \times 10^{-2}$ and $K_{30 °C} = 1.78 \times 10^{-2}$ (mol/liter/mm Brodie) according to Warburg [19].

Some KHCO$_3$/K$_2$CO$_3$ ratios for establishing different CO$_2$ concentrations in the culture atmosphere are given in Table 1.

Table 1. Establishment of high CO$_2$ concentrations by bicarbonate/carbonate buffer mixtures

Buffer mixtures: ml 2M KHCO$_3$ / ml 2M K$_2$CO$_3$	CO$_2$ concentration* % (w/v)
78/22	2.0
60/40	0.7
45/55	0.5
40/60	0.25
30/70	0.13
20/80	0.05

* CO$_2$ concentration in the gaseous atmosphere above the buffer solution in a closed glass flask has been measured using an infrared gas analyser (Finor, Maihak, Hamburg, Germany).

Light

White light from fluorescent tubes (Osram 36W/11, day light; Philips 36W/84) or from mercury vapor lamps (Osram HQLR de Luxe, 85W) is used. Photon flux density of the photosynthetic active radiation (PAR, 400–700 nm) ranging between 100–300 μmol m^{-2} s^{-1} are used. Both continuous illumination as well as light/dark cycles (for example: 16 h light; 8 h darkness) can be used. As continuous illumination supplies more energy to the cells, it will possibly reduce loss of carbon during dark respiration and therefore may be more beneficial to achieve higher growth rates of *in vitro* cultured cells. On the other hand, light-dark-changes will allow 'normal' metabolic processes as they occur in the intact plant.

Culture vessels used for photoautotrophic cell culture growth

A number of culture vessels different in size and shape, such as multiwell dishes, small conical flasks (50–200 ml), two-tier culture vessels, bubble tubes or airlift fermenter systems are used to grow photoautotrophic cell cultures.

A. Closed culture systems

Multi-well dishes for selecting and growing callus cells under photoautotrophic conditions

Fig. 1. Multiwell dish for growing photosynthetically active callus cells under elevated CO_2 concentrations

Steps in the procedure
1. Use presterilized transparent flat bottom plates with lid, containing 6 wells (well diameter 35 mm; depth 17 mm) to select and/or maintain callus cells capable of photoautotrophic growth (Fig. 1).
2. In three or four wells place 5 ml sterile 2 M $KHCO_3$/2 M K_2CO_3 buffer mixtures to establish elevated CO_2 concentrations (0.05%–2% CO_2, v/v) in the gas phase of the dishes.
3. In the remaining wells place 5 ml nutrient agar which contains no or reduced amounts of sucrose.
4. As an inoculum, plate out 0.5–1.0 g cells on nutrient agar. Selection

and/or maintenance of callus cells capable of photoautotrophic growth occurs under sugar famine at elevated CO_2 concentrations.

5. Seal the plates tightly with Parafilm to prevent loss of CO_2 and desiccation and put under continuous or intermittent white light $(100-120 \ \mu mol \ m^{-2} s^{-1})$ at $26 \pm \ °C$.

Notes

1. Presterilized multi-well plates can be purchased from Sigma Chemical Company, Deisenhofen, Germany as well as from C.A. Greiner, Labortechnik, Nürtingen, Germany.

3. Composition of the nutrient agar: The appropriate sugar-free culture medium solidified with 0.9% Difco-agar.

The two-tier culture vessel

The two-tier culture vessel has been developed for the photoautotrophic growth of cell suspensions of *Chenopodium rubrum* [9, 10]. In the author's laboratory, this culture method has been used successfully for the propagation of photoautotrophic cell cultures from 10 different plant species.

Fig. 2. The two-tier culture vessel.

The two-tier culture vessel (Fig. 2) is constructed of two 200 ml small-necked Erlenmeyer flasks, that are connected top-to-bottom, via taper joints 29/32. The upper flask contains a central glass tube (2.0 cm inner diameter; 3.5 cm length within the flask; 3.5 cm length outside the flask including taper joint 29/32. The lower flask contains 50 ml of a 2 M $KHCO_3$/2 M K_2CO_3 buffer solution for establishing the desired CO_2 concentrations in the culture atmosphere. The liberated CO_2 passes through the central glass tube into the upper compartment, that serves as the culture vessel, for propagating the cells in 30 ml sugar-free culture medium.

The two-tier culture flasks (the openings closed by aluminium foil), the culture medium (30 ml), and the bicarbonate/carbonate-buffer solution are separately sterilized by autoclaving for 20 min at 120 °C. The buffer mixture must be filled into a tightly screw-capped glass flask (to prevent loss of CO_2) made of Duran glass (Schott, Germany) that will withhold the pressure during autoclaving.

For routine subculturing 1–2 g cells are inoculated in 30 ml culture medium. For transferring the cells, a stainless steel spoon or pipette may be used.

Sufficient CO_2 (initially 2% CO_2, v/v) is produced by 50 ml of a 2 M $KHCO_3$/2 M K_2CO_{32}-buffer to allow sustained photoautotrophic cell culture growth (increase in fresh weight up to 500%) for 2–3 weeks. At the end of a 14 days subculture period the CO_2 content in the culture atmosphere has changed from initially 2% (v/v) CO_2 to approximately 0.5% (v/v) CO_2, still sufficiently high to support photoautotrophic cell culture growth.

The openings of the culture flask are tightly sealed with sterile aluminium foil and finally with Parafilm to reduce gas exchange with the atmosphere. Using the two-tier culture vessel system, photoautotrophic cell suspensions of *Chenopodium rubrum* are propagated on a gyrotory shaker (120 rpm) illuminated with continuous white light (100–120 μmol m^{-2} s^{-1}) at 26 $\pm$ 1 °C.

B. Open culture systems

The composition of the gas phase developed in the culture vessels may severely affect cell growth and chlorophyll formation. In cell suspension from *Spinacia oleracea* ethylene accumulation and oxygen concentrations above air saturation inhibited greening [2, 4, 13]. Reducing the oxygen content to about 20% of air saturation (photorespiration is reduced without inhibiting mitochondrial respiration) and increasing the CO_2 concentrations to abolish the inhibitory effect of ethylene, favoured sustained chlorophyll formation. In cases where the cultured cells are susceptible to enriched oxygen levels, ethylene or other volatile metabolic compounds, that may accumulate in the closed two-tier culture vessel, it will be necessary or at least beneficial to grow the cells in an open culture system.

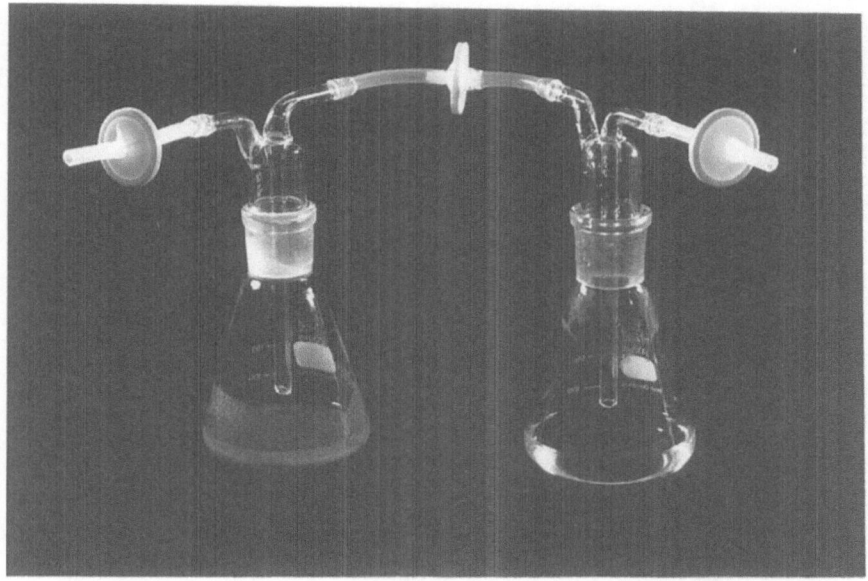

Fig. 3. An open culture system for aerating cells with CO_2-enriched air.

Steps in the procedure
1. Suspend the cells in 50 ml liquid culture medium or grow as callus masses on a nutrient agar in 200 ml Erlenmeyer flasks (Fig. 3).
2. Continuously flush with CO_2-enriched air (1%–2% CO_2, v/v) or with a gas mixture of known composition (1%–2% CO_2, v/v; reduced oxygen levels, 5%–10% O_2, v/v) at 26 ± 1 °C.
3. Keep the gas mixture aseptic and wetted by passing a sterile filter (Milipore filter unit, 0.45 μm pore size) and a water-filled glass flask (humidifier) before entering the culture vessel via small glass tubes.
4. Attach 2 flasks to each humidifier by branching the tubings.
5. Keep the cell cultures under white fluorescent light (continuous or intermittent illumination; 100–120 μmol m^{-2} s^{-1}) at 26 ± 1 °C.
6. Agitate cell suspensions at 120 rpm on a gyrotory shaker (Certomat R, B. Braun Biotech International, Melsungen, Germany).

This culture technique as has been applied successfully for photoautotrophic cell cultures of *Spinacea oleracea* [4], *Nicotiana tabacum* [22] and *Glycine max* [7].

Culture vessel for photoautotrophic cell culture growth under ambient air

The reasons that cultured plant cells in general require highly elevated CO_2 concentrations far above ambient air level for sustained photoautotrophic growth are still unknown. Meanwhile, cell cultures from *Arachis hypogaea, Euphorbia characias, Dianthus caryophyllous* and *Chenopodium rubrum* can be grown photoautotrophically under ambient CO_2 concentrations but at drastically reduced growth rates compared to cells growing under high CO_2.

For photoautotrophic cell culture growth under ambient air the culture flask (200 ml Erlenmeyer flask) is closed by silicon sponge closures (Fig. 4).

Fig. 4. Culture vessel for photoautotrophic cell culture growth under ambient air.

These closures (silicon rubber rings with sealed-in silicone sponge filters) fit well around the neck of 200 ml Erlenmeyer flasks. They permit efficient exchange of gases, but reduce evaporation and prevent passage of aerosols, are reusable and autoclavable and are available from Sigma-Techware, Sigma Chemical Company, St. Louis, U.S.A.

Using this culture technique as described, photoautotrophic cell suspensions from *Chenopodium rubrum* have been grown under atmospheric air for several years.

Determination of photoautotrophic growth parameters

Cell culture growth is monitored throughout the growth cycle by measuring increase in fresh biomass, cell number, packed cell volume, accumulation of chlorophyll, photosynthetic oxygen development and respiratory oxygen consumption.

Procedures

Fresh weight
Determine cell fresh weight after collecting the cells on a fiberglass filter under vacuum.

Packed cell volume
Sediment cells from 1 ml aliquots of the cell suspension by centrifugation (swinging buckets; $g = 200$; 10 min) using calibrated conical glass tubes (Schlee, Witten, Germany).

Cell number
Steps in the procedure
1. Suspend 40 mg cells (fresh weight) in 5 ml chromic acid (10%, w/v) and incubate at 70 °C for 5–10 min.
2. Disperse the cells by drawing the suspension through a needle (1 mm; 40 mm) attached to a syringe.
3. Count the cells in a Fuchs-Rosenthal counting chamber (depth: 0.2 mm) under an inverted microscope (Leitz-Diavert, Leitz, Wetzlar, Germany).

Chlorophyll content
Steps in the procedure
1. Weigh out 200 mg fresh cell biomass into 2 ml round-bottom plastic centrifugation tubes, rapidly freeze, resuspend and extract chlorophyll from the cells with 80% acetone (v/v) by stirring on a magnetic stirrer using magnetic stirring bars.
2. Centrifuge the tubes and pour the clear chlorophyllous supernatant into a graduated glass tube.
3. Repeat this procedure until the pellet is completely free of chlorophyll.
4. Measure absorbance of the chlorophyllous acetone extract in a spectrophotometer at 647 nm and 664 nm and calculate concentration of chlorophyll a and b according to the formula of Ziegler and Egle [24]:
$C_{Chl.a} = 11.78 \times E_{647} - 2.29 \times E_{664}$ (μg chlorophyll a/ml acetone extract);

$C_{Chl.b} = 20.05 \times E_{664} - 4.77 \times E_{647}$ (μg chlorophyll b/ml acetone extract).

Measurements of photosynthetic and respiratory activities
Steps in the procedure
Photosynthetic oxygen production and respiratory oxygen consumption are measured using a Hansatech oxygen meter, equipped with a Clark-type electrode purchased from Bachofer, Reutlingen, Germany.

1. Resuspend 100 mg cells in 1 ml of the same medium in which they had been grown before, supplemented with 20 μM KHCO$_3$, and place into the electrode chamber.
2. After 1 min preincubation, determine dark respiratory oxygen consumption and finally photosynthetic oxygen production in the light (600 μmol m^{-2} s^{-1}).

References

1. Brulfert J, Mricha A, Sossoutov L, Queiroz O (1987) CAM induction by photoperiodism in green callus cultures from a CAM plant. Plant, Cell and Environment 10: 443–449.
2. Dalton C (1980) Photoautotrophy of spinach cells in continuous culture: Photosynthetic development and sustained photoautotrophic growth. J Exp Bot 31: 791–804.
3. Dalton CC (1984) The effect of sugar supply rate on photosynthetic development of *Ocimum basilicum* (sweet basil) cells in continuous culture. J Exp Bot 35: 505–516.
4. Dalton CC, Street HE (1976) The role of the gas phase in the greening and growth of illuminated cell suspension cultures of spinach (*Spinacia oleracea*). In Vitro 12: 485–493.
5. Gamborg O, Miller RA, Ojima K (1968) Nutrient requirements of suspension cultures of soybean root cells. Exp Cell Res 50: 151–158.
6. Gross U, Gilles F, Bender L, Berghöfer P, Neumann KH (1993) The influence of sucrose and elevated CO_2 concentration on photosynthesis of photoautotrophic peanut (*Arachis hypogaea* L.) cell cultures. Plant Cell Tiss Org Cult 33: 143–150.
7. Horn ME, Sherrard J, Widholm JM (1983) Photoautotrophic growth of soybean cells in suspension culture. Plant Physiol 72: 426–429.
8. Hüsemann W (1985) Photoautotrophic growth of cells in culture. In: Vasil IK (Ed.), Cell Culture and Somatic Cell Genetics of Plants, Vol II, Cell Growth, Nutrition, Cytodifferentiation, and Cryopreservation, pp. 213–252, Academic Press, New York.
9. Hüsemann W (1984) Photoautotrophic cell cultures. In: Vasil IK (Ed.) Cell Culture and Somatic Cell Genetics of Plants, Vol I, Laboratory Procedures and their Applications, pp. 182–191, Academic Press, New York.
10. Hüsemann W, Barz W (1977) Photoautotrophic growth and photosynthesis in cell suspension cultures of *Chenopodium rubrum*. Physiol Plant 40: 77–81.
11. Hüsemann W, Fischer K, Mittelbach I, Hübner S, Richter G, Barz W (1989) Photoautotrophic plant cell cultures for studies on primary and secondary metabolism. In: Kurz WGW (Ed.) Primary and Secondary Metabolism of Plant Cell Cultures II, pp. 35–46, Springer-Verlag, Berlin, Heidelberg.
12. Kluge M, Hell R, Pfeffer A, Kramer D (1987) Structural and metabolic properties of green tissue cultures from a CAM plant, *Kalanchoe blossfeldiana, hybr. Montezuma*. Plant, Cell and Environment 10: 451–462.
13. Laulhere JP, Aguettaz P, Lescure A (1984) Regulation of the oxygen exchanges and the greening by controlled supplies of sugar in photomixotrophic spinach cell suspensions. Physiol Veg 22: 765–773.
14. Murashige T, Skoog F (1962) A revised medium for rapid growth and bioassays with tobacco tissue cultures. Physiol Plant 18: 473–497.
15. Rebeille F (1988) Photosynthesis and respiration in air-grown and CO_2-grown photoautotrophic cell suspension of carnation. Plant Sci 54: 11–21.
16. Richter G, Hundrieser J, Groß M, Schultz S, Bottländer K, Schneider Ch (1984) Blue light effects in cell cultures. In: Senger H (Ed) Blue Light Effects in Biological Systems, pp. 387–396, Springer-Verlag, Berlin, Heidelberg.
17. Richter G, Dudel A, Einspanier R, Dannhauer I, Hüsemann W (1987) Blue-light control of mRNA level and transcription during chloroplast differentiation in photomixotrophic and photoautotrophic cell cultures (*Chenopodium rubrum* L.). Planta 172: 79–87.
18. Stöcker S, Guitton M-Ch, Barth A, Mühlbach HP (1993) Photosynthetically active suspension cultures of potato spindle tuber viroid infected tomato cells as tools for studying viroid-host cell interaction. Plant Cell Reports 12: 597–602.
19. Warburg O, Geissler AW, Lorenz S (1962) Neue Methode zur Bestimmung der Kohlensäuredrücke über Bicarbonat/Carbonatgemischen. In: Warburg O (Ed.) Weiterenwicklung der zellphysiologischen Methoden, pp. 578–581, Thieme-Verlag, Stuttgart.
20. Widholm JM (1992) Properties and uses of photoautotrophic plant cell cultures. In: International Review of Cytology, Vol 132, pp. 109–175, Academic Press, New York.

21. Xu Ch, Blair L, Rogers SMD, Godvindjee K, Widholm JM (1988) Characteristics of five new photoautotrophic suspension cultures including two *Amaranthus* species and a cotten strain growing on ambient co_2 levels. Plant Physiol 88: 1297–1302.
22. Yamada Y, Imaizumi K, Sato F, Yasuda T (1981) Photoautotrophic and photomixotrophic culture of green tobacco cells in a jar-fermenter. Plant Cell Physiol 22: 917–922.
23. Yasuda T, Hashimoto T, Sato F, Yamada Y (1980) An efficient method of selecting photoautotrophic cells from cultured heterogenous cells. Plant Cell Physiol 21: 929–932.
24. Ziegler P, Egle K (1965) Zur quantitativen Analyse der Chloroplastenpigmente. Beitr Biol Pflanzen 41: 11–37.

Plant Tissue Culture Manual **H2**, 1–15, 1995.

Zinnia mesophyll culture system to study xylogenesis

MUNETAKA SUGIYAMA & HIROO FUKUDA
Biological Institute, Faculty of Science, Tohoku University, Aoba-yama, Sendai 980–77, Japan

Introduction

Xylem cells are formed from the procambium of the root and shoot in the primary xylem and from the vascular cambium in the secondary xylem. The main components of xylem cells are tracheary elements, which are characterized by the formation of secondary cell walls that thicken with annular, spiral, reticulate or pitted patterns. At maturity, differentiating tracheary elements lose their nuclei and cell contents, leaving behind hollow tubes that form vessels and tracheids. *In vitro*, tracheary elements can be induced from the parenchymatous cells of various plant species by wounding and/or the application of phytohormones [1].

Fukuda and Komamine [2] established an *in vitro* experimental system in which single mesophyll cells of *Zinnia elegans* can redifferentiate directly into tracheary elements without cell division, based on the work of Kohlenbach and Schmidt [3]. The *Zinnia* system is considered to be very well suited for studies of redifferentiation into tracheary elements, because it possesses various characteristic features, which include a high frequency of redifferentiation and a high degree of synchrony (Table 1). A number of new physiological, biochemical and molecular markers, by which the physiological stages of the redifferentiation can be defined, have been found in the *Zinnia* system [1, 4, 5, 6]. Expression of some of these markers has been demonstrated to occur in close association with xylem development *in situ*, as well as *in vitro* [7]. Thus, the *Zinnia* cell system has contributed greatly to the study of xylogenesis. It is anticipated that this system will be employed more extensively in the future. Therefore, we describe here the basic method for culture of isolated *Zinnia* mesophyll cells, together with cytological methods for monitoring the process of differentiation of mesophyll cells into tracheary elements [2, 8].

Table 1. Merits of the *Zinnia* system

---

1. Single-cell system
   A) Little cell-cell interaction
   B) Homogeneous stimulation
   C) Ability to follow visually the sequence of cytodifferentiation
2. Homogeneous starting materials that are previously uncommitted to cytodifferentiation
3. High synchrony and frequency of cytodifferentiation
4. Strict hormonal regulation

Plant material

The first true leaves of young seedlings of *Zinnia elegans* L. are used as the source of mesophyll cells that are to be cultured *in vitro*. The use of healthy seedlings is essential for obtaining viable cells from leaves. Therefore, *Zinnia* seedlings should be grown with great care.

Steps in the procedure

1. Sow seeds of *Zinnia elegans* cv. Canary bird or Envy in moisten vermiculite.
2. Grow seedlings at 25 °C for 14 d under a cycle of 14 h of light (approx. 100 μmol/m²/s, white light from fluorescent lamps) and 10 h of darkness.
3. Water seedlings daily or every other day and feed with 150-fold-diluted HYPONeX (Murakamibussan, Tokyo, Japan) once in the first 5 d.

Notes

(Numbers refer to steps listed above)
1. Seeds of other varieties of *Zinnia elegans* can be used, although there is a slight variation among the rates of differentiation in mesophyll cells isolated from different varieties of seedlings.
2. White fluorescent light should be used. The use of other types of light, such as that from a xenon lamp, sometimes results in failure in the isolation of viable cells, probably because such light induces a high level of phenolics in leaves. Vermiculite should not be allowed to dry up. To avoid bacterial contamination of subsequent cell cultures, it is important to maintain low humidity (approx. 50%) during the growth of seedlings.
3. Water should not come in contact with leaves. Wet leaves are very often associated with bacterial contamination in the subsequent cell culture.

Isolation and culture of mesophyll cells

Since attachment between cells is weak in mesophyll tissues of *Zinnia elegans*, single mesophyll cells can easily be isolated by mechanical maceration of leaves. Steps in the procedure for the isolation and culture of mesophyll cells are described below and are summarized in Fig. 1.

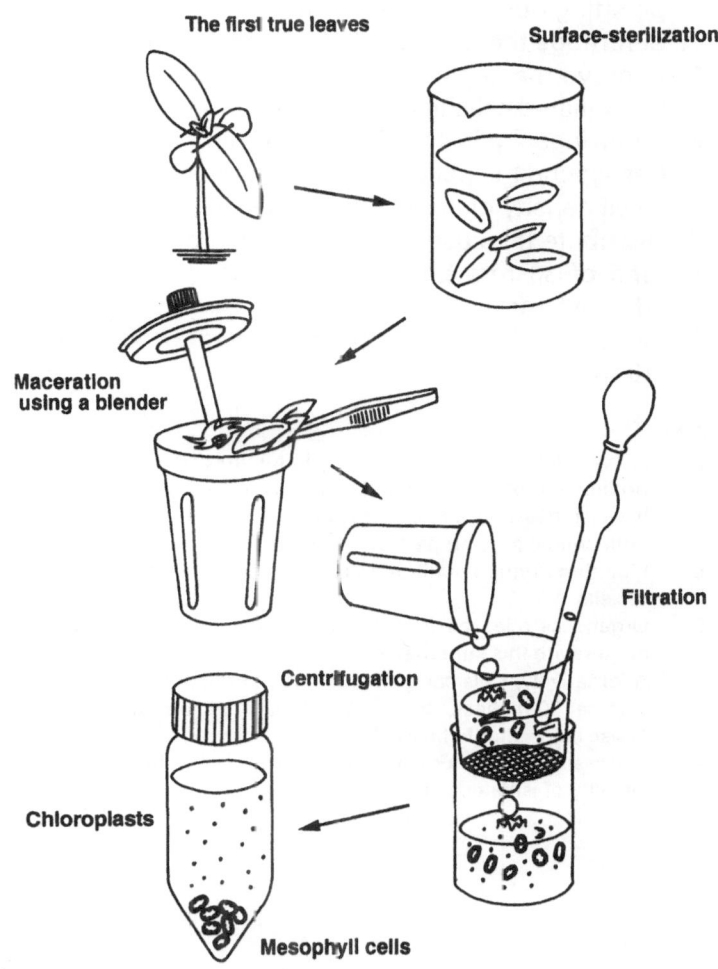

Fig. 1. Procedure for the isolation of mesophyll cells of *Zinnia elegans*.

Steps in the procedure
1. Harvest the first true leaves (60–80 leaves) that are 3 to 4 cm in length.
2. Surface-sterilize the leaves by soaking them for 10 min in 1 liter of

a solution of 0.15% sodium hypochlorite supplemented with 0.001% Triton X-100. Immerse floating leaves occasionally.

3. Rinse the leaves three times with autoclaved, distilled water.
4. Transfer the leaves to the 100 ml stainless-steel cup of a Waring-type homoblender containing 60–80 ml of culture medium.
5. Macerate the leaves at 8,000 rpm for 30 s.
6. Filter the homogenate through a nylon screen with a pore size of 50 to 80 μm. Agitate the homogenate on the screen by repeated pipetting during this filtration.
7. Centrifuge the filtrate at 200 × g for 1 min.
8. Remove the supernatant with a pipette and discard it. Suspend the pelleted cells in 80 ml of medium.
9. Centrifuge again at 200 × g for 1 min.
10. Resuspend the pelleted cells in medium (normally 300–500 ml) at a cell density of about 5 × 10^4 cells/ml.
11. Distribute the suspension of cells into culture tubes (20 ml for a tube of 30 mm i.d. × 200 mm and 3 ml for a tube of 18 mm i.d. × 180 mm).
12. Incubate culture tubes in darkness at 27 °C on a revolving drum (10 rpm).

Notes

2. Leaves should be handled gently to minimize damage to tissues.
3. Sodium hypochlorite should be removed completely.
5. It is also possible to release mesophyll cells by grinding the leaves gently in medium with a smooth pestle and mortar.
6. Pipetting during the filtration is a very effective method for increasing the yield of cells.
8. Organelles released from broken cells, such as chloroplasts, are removed together with this supernatant.
12. Alternatively, cells can be incubated in Erlenmeyer flasks (4 ml of cell suspension per 50-ml flask) on a gyrotary shaker at 75 rpm [9].
2–11. These steps should be carried out on a clean bench.
6–10. These steps should be carried out quickly. Delay during these steps results in low viability of isolated cells.

Culture medium

Table 2 shows the composition of the culture medium for induction of tracheary-element differentiation. This medium is the original medium reported by Fukuda and Komamine [2]. Various modified versions of this medium have been reported [e.g., 10,11]. Medium is prepared by standard method, and sterilized by autoclaving or filtration. Stock solutions for this medium are shown in Table 3. Media without 6-benzyladenine (BA) and/or 1-naphthaleneacetic acid (NAA) can be used for control cultures in which tracheary-element differentiation does not occur.

Table 2. Medium for the culture of *Zinnia* mesophyll cells

Constituents	Concentration (mg liter^{-1})
Macroelements	
KNO_3	2,020
NH_4Cl	54
$MgSO_4 \cdot 7H_2O$	247
$CaCl_2$	147
KH_2PO_4	68
Microelements	
$MnSO_4 \cdot 4H_2O$	25
h_3BO_3	10
$ZnSO_4 \cdot 7H_2O$	10
$Na_2MoO_4 \cdot 2H_2O$	0.25
$CuSO_4 \cdot 5H_2O$	0.025
Na_2EDTA	37
$FeSO_4 \cdot 7H_2O$	28
Organic growth factors	
Glycine	2
myo-Inositol	100
Nicotinic acid	5
Pyridoxine hydrochloride	0.5
Thiamine hydrochloride	0.5
Biotin	0.05
Folic acid	0.5
Phytohormones	
1-Naphthaleneacetic acid (NAA)	0.1
6-Benzyladenine (BA)	1
Sucrose	10,000
D-Mannitol	36,400
pH	5.5

Table 3. Stock solutions for preparation of the medium

Stock A (10×)

KNO_3	20,200 mg
NH_4Cl	540
$MgSO_4 \cdot 7H_2O$	2,470
$CaCl_2 \cdot 2H_2O$	1,470
KH_2PO_4	680
	1,000 ml

Stock B (400×)

$MnSO_4 \cdot 4H_2O$	2,500 mg
H_3BO_3	1,000
$ZnSO_4 \cdot 7H_2O$	1,000
$Na_2MoO_4 \cdot 2H_2O$	25
$CuSO_4 \cdot 5H_2O$	2.5
	250 ml

Stock C (400×)[a]

Na_2EDTA	3,700 mg
$FeSO_4 \cdot 7H_2O$	2,800
	250 ml

Stock D (400×)

Glycine	200 mg
myo-Inositol	10,000
Nicotinic acid	500
Pyridoxine hydrochloride	50
Thiamine hydrochloride	50
Biotin	5
	250 ml

Stock E (400×)[b]

Folic acid	50 mg
	250 ml

[a] Stirred for at least several hours while being heated (100 °C).
[b] Dissolved by adding a small amount of KOH solution.

Determination of the frequency of tracheary-element differentiation

After 2 d in culture, 30–40% of cells differentiate synchronously into tracheary elements. Tracheary elements can easily be identified from the characteristic patterns of their secondary cell walls, which can be seen under a light microscope. The number of tracheary elements is determined using a haemocytometer without any treatment of cells prior to counting. The number of tracheary elements relative to the total number of cells represents the frequency of differentiation to tracheary elements. During culture of mesophyll cells, cell division occurs independently of, but concurrently with, tracheary-element differentiation. The frequency of cell division can be estimated from the number of septa, determined using a haemocytometer, since all the initial mesophyll cells are single cells.

Cytological techniques for monitoring the process of tracheary-element differentiation

Staining of lignin with phloroglucinol

During the differentiation of mesophyll cells to tracheary elements, lignin, a macromolecule composed of hydroxycinnamyl alcohols as monomeric units, accumulates in the secondary cell walls. Lignified tracheary elements can be detected by staining lignin with phloroglucinol-HCl [12]. Phloroglucinol is dissolved in 20% HCl at about 1% (w/v). A drop of this solution and a drop of cell suspension are mixed on a slide. After a 10-min incubation at room temperature, ligninified cell walls are stained reddish purple.

Staining of nuclei with DAPI

Upon the maturation of tracheary elements, intracellular components, including nuclei, chloroplasts, and mitochondria, are lysed autonomously. This stage of differentiation can be monitored by staining nuclei with various dyes. Among these dyes, DNA-specific fluorescent dyes, such as 4',6-diamidino-2-phenylindole (DAPI) are particularly useful. The procedure for DAPI staining, as described here is based on the method of Kuroiwa et al. [13]. Dye solution contains 1 μg/ml DAPI, 0.25 M sucrose, 0.05% 2-mercaptoethanol, 0.6 mM spermidine, 1 mM EDTA, 0.4 mM phenylmethylsulfonyl fluoride (PMSF), and 10 mM Tris-HCl at pH 7.6. Cell suspension is mixed with equal volumes of a 5% solution of glutaraldehyde and the dye solution. The preparation is ready to be examined after a brief incubation (more than 3 min) at room temperature. Under ultraviolet light, nuclei emit blue-white fluorescence. The number of anuclear tracheary elements relative to the total number of tracheary elements is recorded as an index of the extent of autolytic differentiation.

References

1. Fukuda, H (1992) Tracheary element differentiation as a model system of cell differentiation. Int Rev Cytol 136: 289–332.
2. Fukuda H, Komamine A (1980) Establishment of an experimental system for the tracheary element differentiation from single cells isolated from the mesophyll of *Zinnia elegans*. Plant Physiol 65: 57–60.
3. Kohlenbach HW, Schmidt B (1975) Cytodifferenzierung in Form einer direkten Umwandung isolierter Mesophyll-Zellen zu Tracheiden Z Pflanzenphysiol 75: 369–374.
4. Fukuda, H (1994) Redifferentiation of single mesophyll cells into tracheary elements. Int J Plant Sci 155: 262–271.
5. Sugiyama M, Komamine A (1990) Transdifferentiation of quiescent parenchymatous cells into tracheary elements. Cell Differ Dev 31: 77–87.
6. Fukuda H, Yoshimura T, Sato Y, Demura T (1993) In: Komamine A *et al.* (Eds.) Molecular mechanism of xylem differentiation, pp 93–107, J Plant Res Special Issue 3, The Botanical Society of Japan, Tokyo.
7. Demura T, Fukuda H (1994) Novel vascular cell-specific genes whose expression is regulated temporally and spatially during vascular system development. Plant Cell 6: 967–981.
8. Fukuda H, Komamine A (1982) Lignin synthesis and its related enzymes as markers of tracheary-element differentiation in single cells isolated from the mesophyll of *Zinnia elegans*. Planta 155: 423–430.
9. Church DL, Galston AW (1988) Kinetics of determination in the differentiation of isolated mesophyll cells of *Zinnia elegans* to tracheary elements. Plant Physiol 88: 92–96.
10. Sugiyama M, Fukuda H, Komamine A (1986) Effects of nutrient limitation and gamma-irradiation on tracheary element differentiation and cell division in single mesophyll cells of *Zinnia elegans*. Plant Cell Physiol 27: 601–606.
11. Lin Q, Northcote DH (1990) Expression of phenylalanine ammonia-lyase gene during tracheary-element differentiation from cultured mesophyll cells of *Zinnia elegans* L. Planta 182: 591–598.
12. Siegel, SM (1953) On the biosynthesis of lignins. Physiol Plant 6: 134–139.
13. Kuroiwa T, Suzuki T, Ogawa K, Kawano S (1981) The chloroplast nucleus: distribution, number, size, and shape, and a model for the multiplication of the chloroplast genome during chloroplast development. Plant Cell Physiol 22: 381–396.

Plant Tissue Culture Manual **H3**, 1–31, 1995.

Cell cycle studies: Induction of synchrony in suspension cultures of *Catharanthus roseus* cells

H. KODAMA[1] & A. KOMAMINE[2]

[1]*Department of Biology, Faculty of Science, Kyushu University 33, Fukuoka 812, Japan;* [2]*Department of Chemical and Biological Sciences, Japan Women's University, 2–8–1 Mejirodai, Bunkyo-ku, Tokyo 112, Japan*

Introduction

Synchronous cultures have the potential to be very useful tools for studies of the cell cycle because they can provide homogeneous populations of cells at specific phases of the cell cycle, allowing us to study specific biochemical and molecular biological events during the cell cycle. Synchronous cultures of *Catharanthus roseus* (L.) cells are particularly well suited for such studies since it is easy to achieve high reproducibility of the degree of synchrony with such cultures and procedures for synchronization are straightforward. Various aspects of the molecular events during the cell cycle have been investigated with synchronous cultures of *C. roseus*, and several cell-cycle-dependent genes, such as genes for proliferating-cell nuclear antigen, *cyc07* and cyclins, have been isolated [6].

Synchronization of suspension-cultured cells of higher plants is achieved by the arrest of almost all cells at a specific point in the cell cycle, with subsequent release of cells from growth arrest. The techniques that are used to arrest growth of *C. roseus* cells at the G_1 phase of the cell cycle require only the removal and refeeding of phosphate [1] or auxin [9]. As well as such manipulation of the availability of the phosphate and auxin, removal and resupply of other components of the growth medium, such as sucrose and nitrogen, have also been examined for their ability to induce synchronous cell division. In sucrose-limited cultures, cells with 2C DNA level, namely, cells in the G_1 phase, and cells with 4C DNA level, which correspond to cells in the G_2/M phase, accumulate (Fig. 1), as previously reported in suspension cultures of *Acer pseudoplatanus* [3]. However, the viability of cells in sucrose-starved cultures declines rapidly after cell proliferation has ceased [5]. Thus, manipulation of the availability of sucrose is unsuitable for the synchronization of *C. roseus* cells. The accumulation of cells with 2C DNA level can be observed in nitrogen-limited cultures of *C. roseus*, as well as in phosphate-starved cultures (Fig. 1). When nitrogen is added to nitrogen-limited cells, a step-wise increase in cell number is observed (Fig. 2). However, the duration of the first cell cycle (about 45 hours), namely, the period between the addition of nitrogen and the first cell division, is very much longer than that of the second cell cycle (about 35 hours), the period between the first cell division and the second cell division. The additional time required for the first cell cycle may correspond to the time required for the entry of cells from the so-called G_0 state to the G. phase. The proportion of cells that

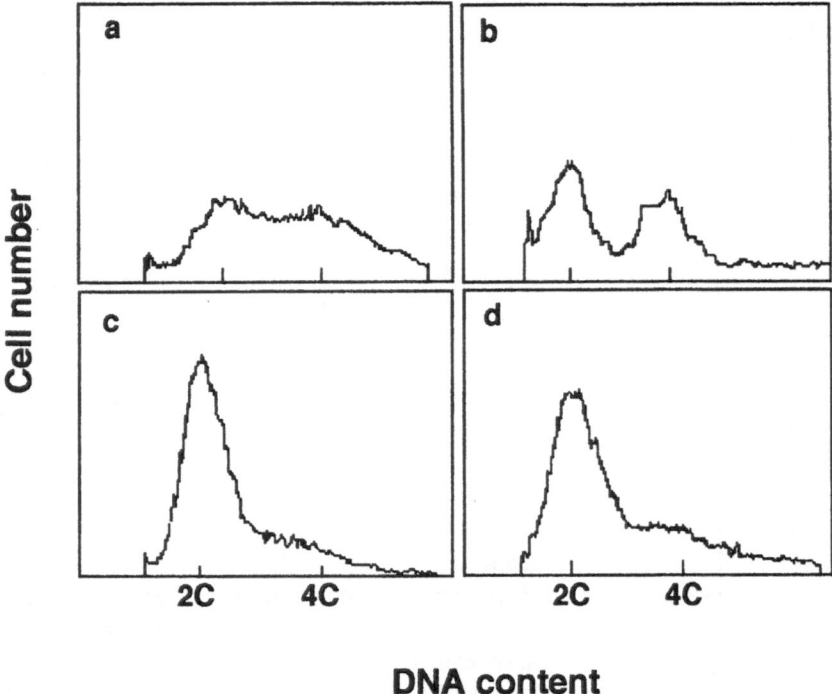

DNA content

Fig. 1. Cell cycle analysis of nutrient-starved cultures by flow cytometry. The gross alterations in the distribution throughout the cell cycle of cells arrested under particular nutrient-starvation conditions were characterized in strain A by flow cytometry. Flow-cytometric analyses are shown of acridine orange-stained protoplasts from cells at the logarithmic phase (a), from a sucrose-limited culture (b), from a nitrogen-limited culture (c), and from a phosphate-starved culture (d). In the sucrose-limited culture, cells were subcultured in sucrose-limited medium that contained about 0.5% (w/v) sucrose. In the nitrogen-limited culture, cells were subcultured in nitrogen-limited medium that contained 1.2 mM nitrogen. Standard MS medium contains 3% (w/v) sucrose and 60 mM nitrogen. Growth of almost all cells ceased 4 days after subculture in the sucrose- or nitrogen-limited medium, and the distribution of cells throughout the cell cycle was determined by flow cytometry 4 days (sucrose-limited culture) and 5 days (nitrogen-limited culture) after subculture. In the phosphate-starved culture, cells were arrested at the G_1 phase by the phosphate starvation method, and flow-cytometric analysis was performed with protoplasts prepared from cells just before the second phosphate feeding of phosphate (see also text for details of steps in synchronization by the phosphate starvation method).

divide during one round of the cell cycle in such cultures is relatively low (about 30 to 50%) and it takes about 10 hours for cells to divide. These properties of cultures subjected to starvation and subsequent addition of nitrogen indicate that the synchrony obtained in such cultures is inadequate for investigations of particular events during the cell cycle.

In synchronous cultures induced by manipulating the availability of phosphate [1] or auxin [9], progression of the cell cycle is rapidly reinitiated, without any delay, after the addition of phosphate or auxin. Cell numbers usually increase by

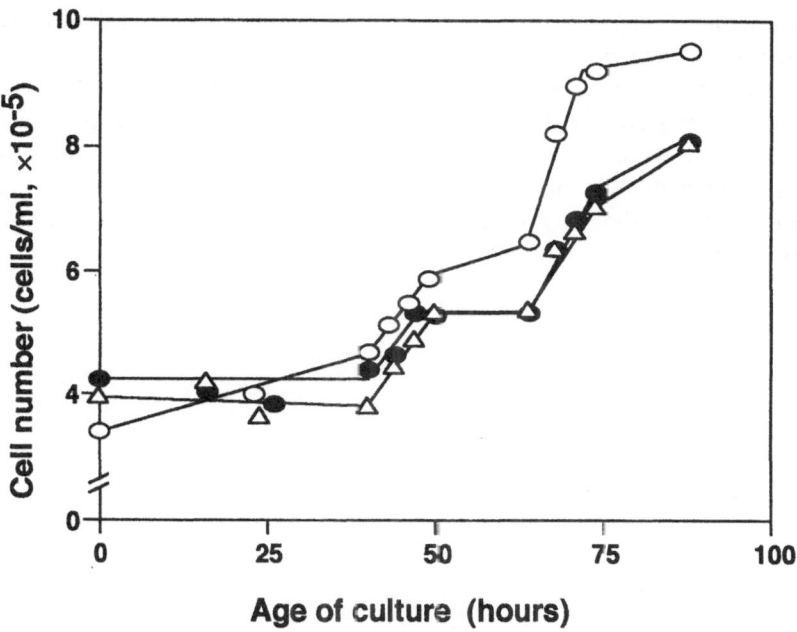

Fig. 2. Changes in cell number after the manipulation of levels of nitrogen in the culture medium. Cells were subcultured in nitrogen-limited medium (total nitrogen, 1.2 mM) and proliferation of cells ceased about 3 days after subculture. Then, nitrogen was added directly to the medium at a final concentration of 60 mM on the 3rd (○), 4th (●) and 5th days (△) after subculture. The age of culture refers to the time after the addition of nitrogen.

70 to 80% within 3 to 4 hours. These results indicate that starvation of phosphate or auxin does cause limited damage to cells. In this chapter we provide details of the procedure for the synchronization of suspension cultures of *C. roseus* cells by the phosphate starvation method, as well as the procedure for synchronization by the auxin starvation method.

Although synchronization by the phosphate starvation method had been considered to be a method whose applicability is limited to suspension-cultured cells of *C. roseus*, recent studies have shown that synchronous cell division can be induced by this method in suspension cultures of *Medicago varia* [4] and *Populus alba* [Dr. H. Maki, personal communication]. The composition of the culture medium may be an important factor for establishment of synchrony by this method. All suspension cultures amenable to synchronization by the phosphate starvation method are maintained in Murashige-Skoog (MS) medium [8]. Standard MS medium contains a relatively low concentration of phosphate, in contrast to elevated levels of nitrogen sources. It seems likely that, in this type of medium, the availability of phosphate limits the division of cells. Therefore, when suspension cultures of other higher plant species are to be synchronized, it is possible that, if cells can be maintained in MS medium and the cultures contain homogeneous and finely dispersed cells, a high degree synchrony may be attainable by this method.

Only a few strains appear to be suitable for synchronization by the auxin starvation method. Among three strains of *C. roseus* cells maintained in the authors' laboratory, cells of only one strain (TN21) can be well synchronized by the auxin starvation method. In the case of cells that have been in culture for long periods of time, this method of synchronization may be unsuitable because of habituation of cells to exogenous auxin.

Procedures for determination of percentages of living cells in cultures

A high percentage of actively dividing cells is required if a culture is to be used for successful synchronization because synchrony is strongly influenced by the percentage of non-dividing cells. The proportion of non-dividing cells can be roughly estimated by measuring the viability of cells in a rapidly growing culture (e.g., a culture in the logarithmic phase of growth). Cell viability can be determined by staining the living cells with a 0.12 mM solution of fluorescein diacetate by the method of Widholm [10] and we usually examine at least 500 cells (see also chapter A3, this volume, for details).

Procedures for determination of cell numbers

Cell clusters of all strains of *C. roseus* can easily be macerated by wall-degrading enzymes to yield protoplasts with an efficiency of almost 100%.

Steps in the procedure
1. Transfer three ml of a suspension of cells at a concentration of about 1×10^5 to 5×10^5 cells/ml to a 10-ml test tube, and then pellet cell clusters by centrifugation at $800 \times g$ for 2 min.
2. Remove the supernatant and replace by a mannitol solution.
3. After cell clusters have been suspended in a uniform manner, transfer 2 ml of the suspension to a 10-ml Erlenmeyer flask, and then add 0.5 ml of enzyme solution I (for strain A or B) or II (for strain TN21).
4. Incubate the mixture for 60 min at 27 °C on a shaker operated at 60 strokes/min.
5. Estimate cell numbers by counting protoplasts with a haemocytometer.

Procedures for determination of mitotic indices (Feulgen stain method)

The mitotic index is determined by counting the number of Feulgen-stained nuclei or nuclei stained with 4-6-diamidino-2-phenylindole (DAPI) that are observed to be in mitosis.

Steps in the procedure
1. Mix one volume of cell suspension and five volumes of fixative (10% formaldehyde in 30 mM sodium-phosphate buffer, pH 7.0).
2. After 24 h, wash cells with distilled water and hydrolyze in 1 M HCl for 10 min at 60 °C.
3. Stain in Schiff's reagent for 3 h and wash with 2% (w/v) sodium bisulfite and then with water.
4. Determine the mitotic index in the squash of Feulgen-stained nuclei (more than 1,000) that are clearly recognized to be at metaphase or anaphase in a given population of cells.

Procedures for determination of mitotic indices (DAPI stain method)

Steps in the procedure
1. Transfer 200 μl of a suspension of cells to a 1.5-ml microcentrifuge tube and pellet cell clusters by centrifugation at 3,000 × g for 1 min.
2. Resuspend in 2% (w/v) glutaraldehyde in 30 mM potassium phosphate buffer (pH 7.0). The fixed cells can be stored for several days at 4 °C.
3. Remove the supernatant by centrifugation at 3,000 × g for 1 min and hydrolyze in 1 M HCl for 15 min at 60 °C.
4. Collect cells by centrifugation at 3,000 × g for 1 min and resuspend in 2% (w/v) glutaraldehyde in 30 mM potassium phosphate buffer (pH 7.0). Place approximately 10 μl of a suspension of fixed cells on a slide, mix with an approximately equal volume of a DAPI solution and stain for a few minutes.
5. Determine the mitotic index in the squash of DAPI-stained nuclei (more than 1,000) that are clearly at metaphase or anaphase in a given population of cells. Examine stained cells under an fluorescence microscope fitted with a DApo 100 UV PL objective lens (Olympus Kogaku Kogyo) with fluorescence illumination at 360 nm for excitation of DAPI.

Procedures for determination of incorporation of [³H]thymidine

Steps in the procedure
1. The rate of DNA synthesis is measured by labeling cells with [³H]thymidine (111 kBq/ml suspension, 1.67 TBq/mmol, for example) for 60 min at 27 °C. After incubation, add 1.6 ml of ice-cold ethanol to 0.4 ml of the suspension of labeled cells.
2. Transfer the cells to a 1.5-ml microcentrifuge tube and centrifuge at 2,000 × g for 10 min at 4 °C.
3. Wash the pellet twice with ice-cold 80% (v/v) ethanol and then with ice-cold 0.2 M perchloric acid (PCA).
4. Solubilize nucleic acids by heating in 0.5 M PCA at 80 °C for 15 min.
5. Centrifuge the extract at 10,000 × g for 10 min, and measure an aliquot of radioactivity of the supernatant, after combination with a commercial scintillant, in a liquid scintillation counter.

Procedures for flow cytometry

Analyze the DNA content of protoplasts that have been stained with acridine orange by flow cytometry as described in Ando *et al.* [2].

Steps in the procedure
1. Isolate protoplasts of *C. roseus* as described for the procedures of determination of cell numbers.
2. Collect protoplasts by centrifugation at $500 \times g$ for 3 min, wash once with a mannitol solution and fix in 70% (v/v) ethanol.
3. Treat protoplasts (about 10^6/sample) with 10 ml of a solution containing 0.08 M hydrochloric acid, 0.1% Triton X-100, and 0.08 M NaCl in ice bath for 5 min.
4. Collect protoplasts by centrifugation at $500 \times g$ for 3 min and stain with 12 ml of a solution containing 4 mg/l acridine orange, 1 mM EDTA-2Na, 0.15 M NaCl, and 0.12 M phosphate-citrate buffer (pH 6.0), in an ice bath for about 2 h.
5. Histograms of green fluorescence are obtained from a total 20,000 protoplasts at a flow rate of 100 to 150 cells/second with an flow cytometer fitted with a 100 µm nozzle. The laser output is 480 nm at a 600 mW and barrier filters are LP515 and SP530.

Solutions

Enzyme solution I (for cells of strain A or B)
10% (w/v) Cellulase "Onozuka" R-10 (Yakult Honsha, Tokyo, Japan)
5% (w/v) Macerozyme R-10 (Yakult Honsha, Tokyo, Japan)
 Dissolve both enzymes in H_2O. The solution of enzymes can be stored for several months at -20 °C.

Enzyme solution II (for cells of strain TN21)
10% (w/v) Cellulase "Onozuka" R-10
5% (w/v) Macerozyme R-10
5% (w/v) Pectolyase Y-23 (Seishin Seiyaku, Tokyo, Japan)
 Dissolve all three enzymes in H_2O. Insoluble materials should be removed from the solution by centrifugation at 10,000 × g for 10 min at 4 °C. The supernatant can be stored for several months at -20 °C.

Mannitol solution
0.6 M mannitol containing 1% (w/v) $CaCl_2 \cdot 2H_2O$

DAPI solution
Dissolve DAPI at a final concentration of 1 µg/ml in S buffer [7] which contains 20 mM Tris-HCl (pH 7.6), 0.25 M sucrose, 1 mM EDTA, 0.6 M spermidine, 7 mM β-mercaptoethanol, 1 mM phenylmethylsulphonyl fluoride.

MS medium
Murashige and Skoog basal medium [8]
supplemented with:
2.2 µM 2,4-dichlorophenoxyacetic acid (2,4-D) for cells of strain A (or B), or 4.4 µM 2,4-D for cells of strain TN21, and 3% (w/v) sucrose.
 Adjust the pH to 5.7 with 1 M KOH. In our laboratory, some mineral components, namely, NH_4NO_3, KNO_3, H_3BO_3, $MnSO_4 \cdot 4H_2O$, $ZnSO_4 \cdot 7H_2O$, KI, $Na_2MoO_4 \cdot 2H_2O$, $CoCl_2 \cdot 6H_2O$ and $CuSO_4 \cdot 5H_2O$, are dissolved together as a concentrated stock solution (× 50). Solutions of KH_2PO_4, Fe-EDTA, $CaCl_2 \cdot 2H_2O$ and $MgSO_4 \cdot 7H_2O$ are prepared separately as individual stock solutions (× 200). By preparing the stock solution of KH_2PO_4 separately from those of other mineral components, we can easily prepare phosphate-free MS medium.

Procedures for synchronization of cell division by the phosphate starvation method

Establishment of suspension cultures of strain A (or B) of C. roseus cells

The available laboratory strains of *C. roseus* cells consist of homogeneous and finely dispersed cells. Suspension cultures of strain A (or B) of *C. roseus* (L.) G Don can be synchronized by the phosphate starvation method. These strains were initiated originally from a culture of stem tissue in 1969. Cells are maintained at 27 °C, in darkness, in MS medium that contains 3% (w/v) sucrose and 2.2 μM 2,4-D. Cells are subcultured at 7-day intervals by the transfer of 7 ml of the suspension of cells into 43 ml of fresh medium in a 300-ml Erlenmeyer flask, and flasks are shaken 80 to 90 strokes/min on a shaker.

Concentration of phosphate and proliferation of cells

The growth cycle of suspension-cultured *C. roseus* cells of strain A (or B) consists of a lag phase of about 1 day, a logarithmic phase of 4 days and a stationary phase that is reached about 6 days after subculture.

The growth of *C. roseus* cells in MS medium is limited by the concentration of phosphate in the medium. The relationship between the concentration of phosphate and the proliferation of cells should be determined as follows. Cells at the stationary phase (about 6 to 9 days after subculture) are cultured for 3 days in phosphate-free MS medium at an initial density of about 3×10^5 cells/ml. Then phosphate is added as KH_2PO_4 to cultures to final concentrations of 0 to 1.25 mM. Cells are cultured for a further 9 days (except in the case of the culture with 0 mM phosphate) and the number of cells is counted. The number of cells in the phosphate-free culture is determined 3 days after the addition of phosphate to the other cultures. A linear relationship should be obtained between the concentration of phosphate in the medium (0 to 1.25 mM; standard MS medium contains 1.25 mM phosphate) and the number of cells 9 days after the addition of phosphate. Figure 3 shows that the population of phosphate-starved cells at an initial density of 3×10^5 cells/ml can double in cell number after the addition of phosphate at a final concentration of 0.14 mM.

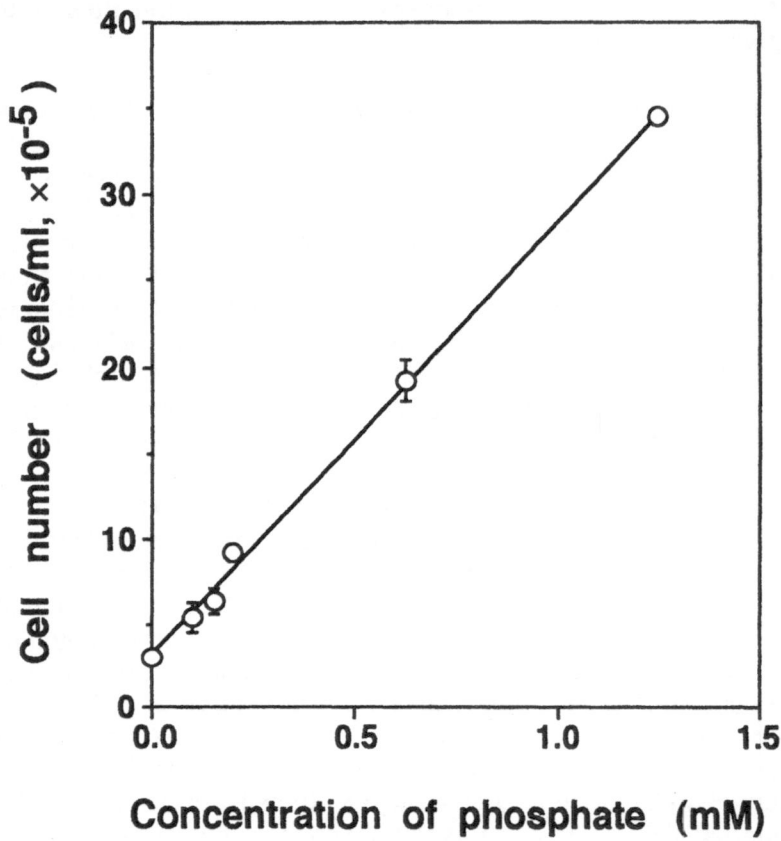

Fig. 3. Effects of the concentration of phosphate on the proliferation of *C. roseus* cells (strain A). Nine-day-cultured cells were transferred to phosphate-free medium. After culture for 3 days, phosphate was added to the medium at various concentrations, as indicated. After proliferation of cells had ceased, cell numbers were determined. Vertical lines indicate standard deviations. Their absence in this and other Figures indicates that standard deviations fell within symbols.

Steps in the procedure

A diagram of the procedure for implementation of the phosphate starvation method is shown in Fig. 4.

1. First phosphate starvation (S1)

During all steps described, cells are cultured at 27 °C in darkness, with shaking at 80 to 90 strokes/min on a shaker. Transfer cells at the stationary phase (about 6 to 9 days after subculture) directly to phosphate-free medium and culture for 3 days in 100 to 150 ml of phosphate-free medium at an initial density of about 3×10^5 cells/ml in a 500-ml Erlenmeyer flask.

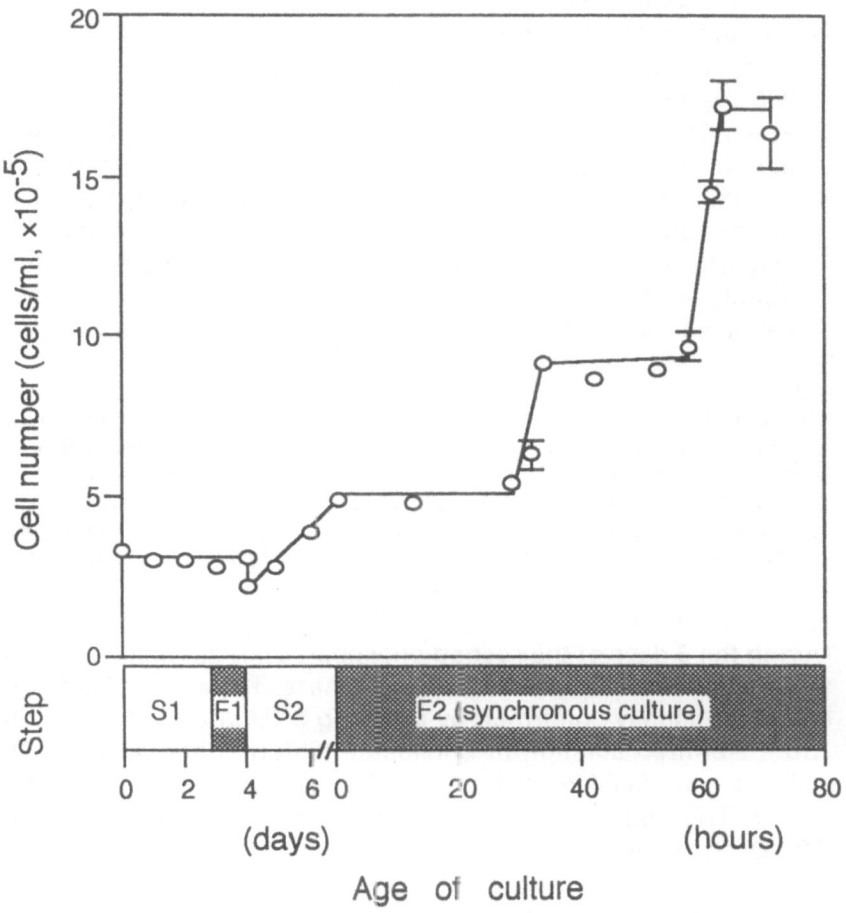

Fig. 4. Diagram of the steps required for synchronization by the phosphate starvation method. The abbreviations, namely, S1, F1, S2 and F2, are explained in the text. Vertical lines indicate standard deviations.

Note
This culture period is designated the first phosphate starvation (S1). It is important that no apparent proliferation of cells be observed during this culture period. If numbers of cells increase during this culture period, cells at later stationary phase, for example, 10 or 11 days after subculture, can be cultured in the phosphate-free medium.

2. First phosphate feeding (F1)
Add a sterilized solution of phosphate (250 mM KH_2PO_4) directly to the medium to a final concentration of 0.14 mM. Cells are cultured for 14 to 18 h. This culture period is designated the first phosphate feeding (F1).

3. Second phosphate starvation (S2)

1. Fourteen to eighteen hours after the start of F1, transfer the cells to a 50-ml plastic centrifuge tube and collect by centrifugation at $800 \times g$ for 2 min.
2. Wash the cells twice with phosphate-free medium and resuspend in phosphate-free medium at a density of about 2×10^5 cells/ml.
3. Mix this suspension of cells gently by stirring with a magnetic stirrer for homogeneous dispersion of cells and then divide into 10 ml aliquots in 50-ml Erlenmeyer flasks. The cell number should double during a 3-day culture under these second phosphate-starvation conditions (abbreviated as S2 in Fig. 4).

Note
It is essential, at this step, that excess phosphate be eliminated initially by washing cells with phosphate-free medium. It is important that the cells be washed before they begin to divide. Cells in the process of mitosis are quite liable to suffer from effects that are unfavorable for subsequent synchronization during the washing step. The number of cells increases gradually within 20 to 24 h after the first phosphate feeding. Therefore, we usually wash cells about 14 to 18 h after the start of the first phosphate feeding.

4. Second phosphate feeding (F2)

During the 3 days of the second phosphate starvation, most cells are arrested preferentially at the early G_1 phase. The arrest of cell proliferation is usually confirmed by counting cells at 8-h intervals. Then add a sterilized solution of phosphate (250 mM KH_2PO_4) to a final concentration of 0.625 mM and synchronized cell division can be observed. This culture period is referred as second phosphate feeding (F2), as shown in Fig. 4.

The degree of synchrony

Synchronous division of cells occurs 27 to 31 h after the second feeding of phosphate. The cell number increases by 70 to 80%. The S phase, which is determined by monitoring incorporation of [³H]thymidine into the DNA fraction, is 6 to 17 h in length after the second feeding of phosphate. A sharp increase in mitotic index is observed 26 to 30 h after the second feeding of phosphate (Fig. 5).

A DNA histogram of protoplasts prepared from the cells in the G_1 phase shows only one peak (peak A), which corresponds to a 2C level of DNA. Almost all protoplasts prepared from cells in the G_2 phase have a 4C level of DNA (peak B). In a histogram of protoplasts from cells in cytokinesis, two peaks are observed and they correspond to 2C (peak C) and 4C (peak D) levels of DNA, respectively. The former seems to correspond to cells after cytokinesis and the latter to cells before cytokinesis (Fig. 6).

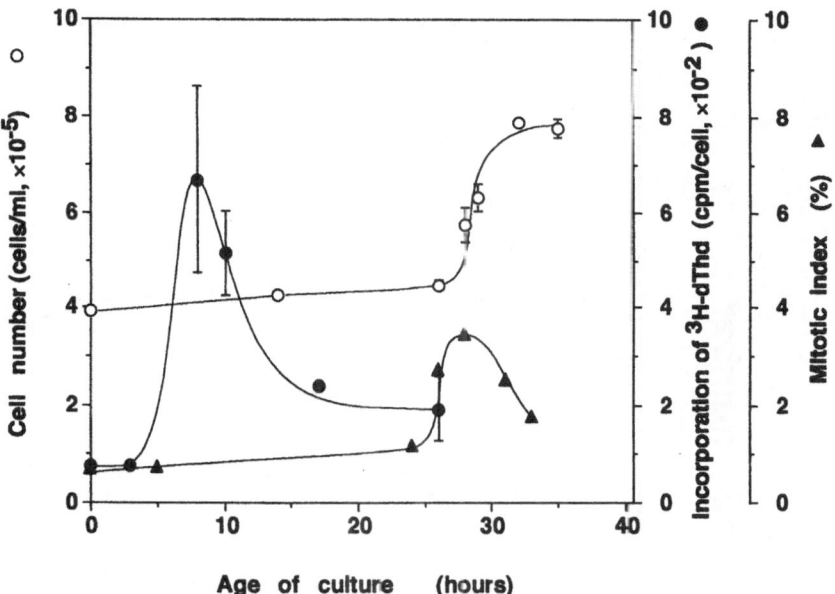

Fig. 5. Changes in cell number, incorporation of [³H]thymidine (³H-dThd) into the DNA fraction and mitotic indices in a synchronous culture of *C. roseus* (strain A), in which synchronization was achieved by the phosphate starvation method. The age of culture refers to the time after the second feeding of phosphate. Vertical lines indicate standard deviations.

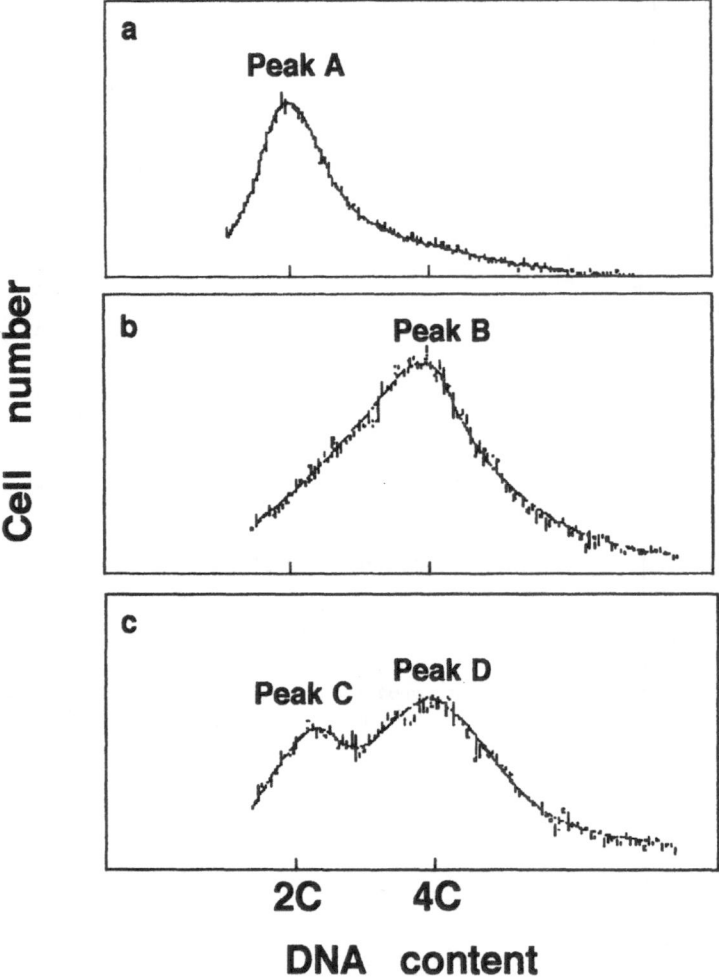

Fig. 6. Distribution of the DNA content in a synchronous culture of *C. roseus* cells in which synchronization was induced by the phosphate starvation method. Flow-cytometric analysis is shown of acridine orange-stained protoplasts from the cells in the G_1 phase (a), in the G_2 phase (b) and during cytokinesis (c). Peaks A through D are discussed in the text.

Procedures for synchronization of cell division by the auxin starvation method

Establishment of suspension cultures of strain TN21 of C. roseus *cells*

Suspension cultures of strain TN21 of *C. roseus* (L.) cv. Little Pinky can be synchronized by the auxin starvation method. Although this strain was originally initiated from a culture of anthers in 1988, almost all cells in cultures are diploid. Cells of strain TN21 are maintained under the same culture conditions as described for the phosphate starvation method (see *establishment of suspension cultures of strain A (or B) of* C. roseus *cells*), with the exception of a concentration of 2,4-D (4.4 μM).

Effects of auxin starvation on the proliferation of cells of the strain TN21

The growth cycle of cells of strain TN21 consists of a lag phase of about 1 day, a logarithmic phase of 5 to 6 days and a stationary phase that is reached after about 8 days of subculture. High cell viability (about 80%) is observed in the culture at the stationary phase.

A unique property of the cells of strain TN21 becomes apparent when cells at the stationary phase are transferred to fresh medium without 2,4-D. After several washes with auxin-free medium, cells are cultured under auxin-starved conditions. No increase in cell number is detectable in auxin-starved cultures of strain TN21 for at least 8 days (Fig. 7). If 2,4-D is added to the medium at a final concentration of 2.2 μM on days zero, 2, and 4 after 2,4-D is eliminated, rapid restoration of cell growth can be observed within 2 days after the addition of 2,4-D (Fig. 7). Thus, high cell viability is clearly maintained in cultures free of auxin for at least 4 days. Synchronization by the auxin starvation method is based on these results: cells of strain TN21 at the stationary phase can be arrested by transfer to fresh, auxin-free medium and refeeding of auxin rapidly induces the cell growth. By contrast, cells of strain A or B, used for synchronization by the phosphate starvation method, increase in number even under auxin-starved conditions. Thus, synchronization by the auxin starvation method can be applied only to cells of strain TN21 in our laboratories. However, the phenomenon of auxin-induced rapid restoration of cell division is very attractive for studies of the role of auxin in the control of cell proliferation. Therefore, it is worth examining the possibility of synchronization by this method in suspension cultures of other plant cells, if cells at the stationary phase show no increase in cell number after transfer to auxin-free fresh medium.

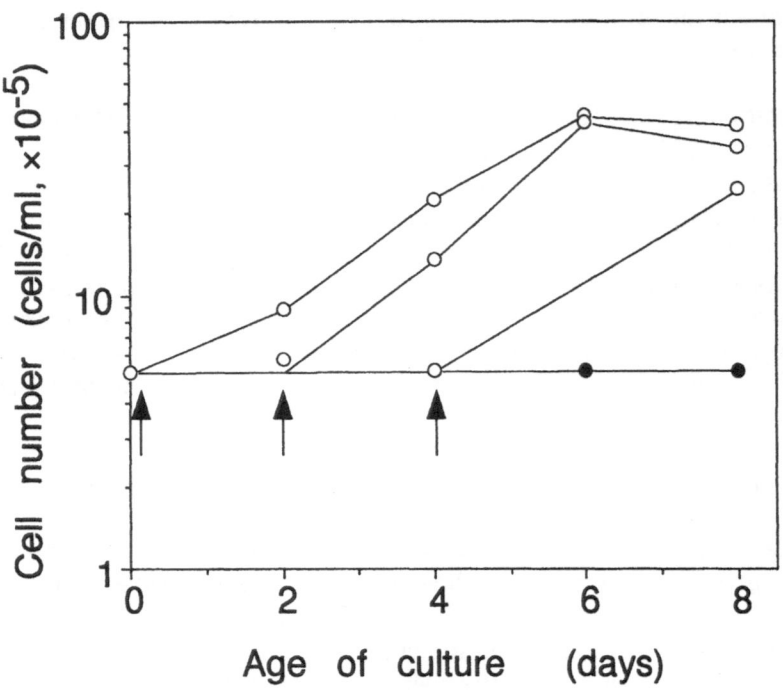

Fig. 7. Effects of 2,4-D on the proliferation of *C. roseus* cells (strain TN21) after auxin starvation. Changes are shown in the numbers of cells in auxin-free medium (•) and in medium supplemented with auxin (○). The age of the culture refers to the time after the transfer of cells to auxin-free medium. Arrows indicate the time at which 2,4-D was added.

Steps in the procedure
A diagram of the procedure for implementation of the auxin starvation method is shown in Fig. 8.

1. Auxin starvation
1. Transfer eight-day-cultured cells of strain TN21 to a 50-ml plastic centrifuge tube and wash 4 times by centrifugation at $800 \times g$ for 2 min with auxin-free MS medium.
2. Resuspend in auxin-free MS medium at a density of about 5×10^5 cells/ml.
3. Divide the suspension of cells into 10 ml aliquots in 50-ml Erlenmeyer flasks, as described for the phosphate starvation method (step 3), and culture for 2 days.

Note
With strain TN21, the density of cells during auxin starvation is important for the maintenance of high cell viability. A density of cells below 5×10^5 cells/ml often results in a decline in cell viability.

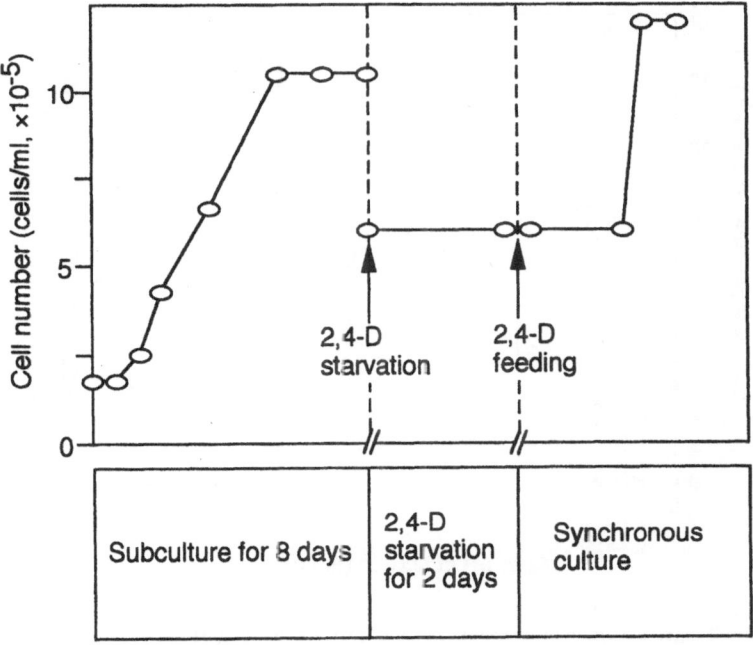

Fig. 8. Diagram of the steps required for synchronization by the auxin starvation method.

2. Auxin feeding
During 2 days of culture with auxin starvation, no apparent proliferation of cells is observed. Add 2,4-D directly to the medium at a final concentration of 4.4 μM to induce synchronized division of cells.

The degree of synchrony
Synchronous division of cells occurs 12 to 15 h after the refeeding of 2,4-D to cultures and cell numbers increase by 75% within 3 to 4 h. A single and clear peak of radioactivity due to incorporation of [³H]thymidine can be observed 9 h after the addition of auxin. The presence of a period of active synthesis of DNA (S phase) indicates that the auxin-starved cells are arrested preferentially at the G_1 phase. The mitotic index reaches a maximum value about 14 h after the addition of auxin, and it also shows a clear peak (Fig. 9).

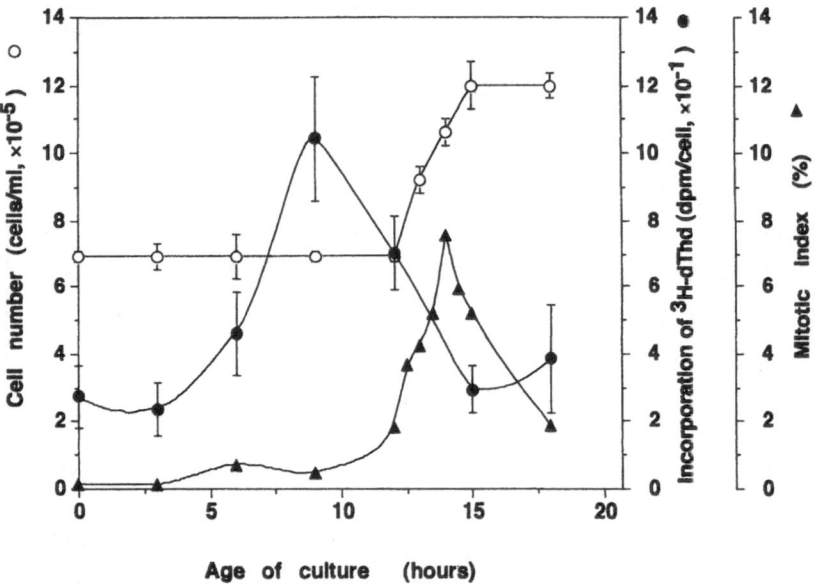

Fig. 9. Changes in cell number, incorporation of [³H] thymidine (³H-dThd) into the DNA fraction and mitotic indices in a synchronous culture of *C. roseus* cells (strain TN21) in which synchronization was induced by the auxin starvation method. The age of the culture refers to the time after the addition of auxin. Vertical lines indicate standard deviations.

Material and suppliers

1. Fluorescein diacetate: Aldrich Chemical Company, Milwaukee, WI, U.S.A.
2. Haemocytometer: Tatai type (or Fuchs-Rosenthal type); Kayagaki Irika Kogyo, Tokyo, Japan.
3. Schiff's reagent: Wako Pure Chemical Industries, Osaka, Japan
4. DAPI: Sigma, St Louis, MO, U.S.A.
5. Fluorescence microscope: Olympus BH-2 fluorescence microscope; Olympus Kogaku Kogyo, Tokyo, Japan.
6. [^{3}H]thymidine: Amersham, Buckinghamshire, England.
7. Liquid scintillation counter: 1216 RACKBETA II; LKB, Finland.
8. Flow cytometer: EPICS V flow cytometer; Coulter Electronics, Hialeah, FL, U.S.A.
9. 50-ml plastic centrifuge tubes: Sumilon centrifuge tubes; Sumitomo Bakelite, Tokyo, Japan.
    These tubes can be autoclaved several times at 121 °C for sterilization.

Acknowledgments

The authors thank Mr. K. Kusumi and Mr. H. Inada (Kyushu University) for their assistance in the preparation of Figures.

References

1. Amino S, Fujimura T, Komamine A (1983) Synchrony induced by double phosphate starvation in a suspension culture of *Catharanthus roseus*. Physiol Plant 59: 393–396
2. Ando S, Shimizu T, Kodama H, Amino S. Komamine A (1987) Flow cytometric analysis of the cell cycle in synchronous culture of *Catharanthus roseus*. Agric Biol Chem 51: 1443–1445
3. Gould AR, Everett NP, Wang TL, Street HE (1981) Studies on the control of the cell cycle in cultured plant cells. I. Effects of nutrient limitation and nutrient starvation. Protoplasma 106: 1–13
4. Kapros T, Bögre L, Németh K, Báko L, Gyögyey J, Wu SC, Dudits D (1992) Differential expression of histone H3 gene variants during cell cycle and somatic embryogenesis in alfalfa. Plant Physiol 98: 621–625
5. Kodama H, Ando S, Komamine A (1990) Detection of mRNAs correlated with proliferation of cells in suspension cultures of *Catharanthus roseus*. Physiol Plant 79: 319–326
6. Kodama H, Komamine A (1994) Studies of the plant cell cycle in synchronous cultures of *Catharanthus roseus* cells. Plant Cell Physiol 35: 529–537
7. Kuroiwa T, Suzuki T, Ogawa K, Kawano S (1981) The chloroplast nucleus: distribution, number, size, and shape, and a model for the multiplication of the chloroplast genome during chloroplast development. Plant Cell Physiol 22: 381–396
8. Murashige T, Skoog F (1962) A revised medium for rapid growth and bio assays with tobacco tissue cultures. Physiol Plant 15: 473–497
9. Nishida T, Ohnishi N, Kodama H, Komamine A (1992) Establishment of synchrony by starvation and readdition of auxin in suspension cultures of *Catharanthus roseus* cells. Plant Cell Tissue Organ Culture 28: 37–43
10. Widholm JM (1972) The use of fluorescein diacetate and phenosafranine for determining viability of cultured plant cells. Stain Technol 47: 189–194

Index